What is Philosophy of Science?

Polity's *What is Philosophy?* series

Sparkling introductions to the key topics in philosophy, written with zero jargon by leading philosophers.

Stephen Hetherington, *What is Epistemology?*

Dean Rickles, *What is Philosophy of Science?*

James P. Sterba, *What is Ethics?*

Charles Taliaferro, *What is Philosophy of Religion?*

What is Philosophy of Science?

Dean Rickles

polity

First published in 2020 by Polity Press

Polity Press
65 Bridge Street
Cambridge CB2 1UR, UK

Polity Press
101 Station Landing
Suite 300
Medford, MA 02155, USA

ISBN-13: 978-1-5095-3416-6
ISBN-13: 978-1-5095-3417-3(pb)

A catalogue record for this book is available from the British Library.

Typeset in 11 on 13 pt Sabon by
Servis Filmsetting Ltd, Stockport, Cheshire
Printed and bound in Great Britain by CPI Group (UK) Ltd, Croydon

For further information on Polity, visit our website: politybooks.com

Contents

Figures and Table

Preface

> [N]o science is possible without a philosophical background.
>
> Ralf Hagedorn

> There is no such thing as philosophy-free science; there is only science whose philosophical baggage is taken on board without examination.
>
> Daniel Dennett

> Science and Philosophy must supplement each other, urge each other forward. Without science, philosophy is null; without philosophy, science is blind.
>
> Antoine Augustin Cournot

Despite these grand proclamations of a necessary union between science and philosophy, philosophy books have an unfortunate tendency to be placed near the "New Age," "Spiritualism," and "Mysticism" shelves of book sellers, far away from the proud, upstanding science books. Yet philosophers of science are usually "scientist-friendly" (many having trained as scientists before "the switch"), though they might view scientists as somewhat naive in their views of how science actually works. Scientists friendly to philosophers are, these days at least, the exception rather than the rule: now, the two subjects, philosophy of science and science itself, are viewed more in opposition than alignment – one can find countless videos on YouTube of Richard Feynman, Neil deGrasse Tyson, and Lawrence Krauss and others bashing philosophy and philosophers; in his last book, the late Stephen Hawking went so far as to say that "philosophy is dead," since it's so out of touch with scientific developments. Ouch ... This was not always the case, and there are in fact signs that things are changing, with new science–philosophy collaborations and some scientists actively encouraging a dialogue with philosophers.

What is philosophy of science? To some extent it is the subject that attempts to provide an answer to the question "what is science?" Philosophy of science puts science itself under the microscope. This book will explain what this looks like. At a simplistic level, science asks why something in the world is so: what makes it *go*? Philosophy of science then asks what makes *science* go: how are its claims *justified*? How, if at all, are its claims distinct from other claims about the world? How is it that the claims it makes about the world can be revised if, as is often suggested, it is supposed to provide *objectively true* descriptions? Can its claims really be said to *map onto* the real world at all? Each of these questions roughly corresponds to the topics covered in the three main chapters of this book (chapters 2–4) – we begin with a broad overview (chapter 1). Each chapter concludes with a brief summary of key concepts followed by an annotated selection of readings.

This book is written for the absolute beginner with no previous exposure to philosophy of science, or science for that matter, to prepare them for more in-depth study. This is, then, intended to be more a *companion* to one of the many other standard textbooks rather than a standalone textbook. It aims to present concisely and in a very simple fashion the key questions and problems defining the subject of philosophy of science, describing also how it has developed as a field and how it links up to broader issues in philosophy.

References

Cournot, A. A., *Essai sur les Fondements de nos Connaissances et sur les Caractères de la Critique Philosophique* (Hachette, 1851).

Dennett, D., *Darwin's Dangerous Idea* (Simon & Schuster, 1995).

Hagedorn, R., "What Happened to Our Elementary Particles?" In C. Enz and J. Mehra (eds.), *Physical Reality and Mathematical Description* (Reidel, 1974), pp. 100–10.

1

Philosophy, Science, and History

I am assuming that the reader I am addressing is an absolute beginner, taking (or thinking of taking) a philosophy of science course for the first time; or perhaps a non-student simply wishing to have a better critical understanding of science. It's no easy task to state exactly what *science* is (indeed, that is one of the chief problems tackled by philosophers of science), and so it is doubly difficult to spell out what one means by "*philosophy of* science." But this book aims to do just that. We begin in this chapter with some, at this stage very loose (and slightly repetitive – to drill some major themes in), remarks about the nature of philosophy, the nature of science, and their union. Then we present the subject through its history, describing, in very broad brushstrokes, the key stages leading to the kinds of issues discussed by the philosophers of science of today. We start by sweeping aside some common misconceptions about the nature of philosophy.

Common Myths About Philosophy

Since many reading this (perhaps *most*, in fact) will not be philosophy students, it might be a good idea to say something about why you should study philosophy of science, and also to dispel some common myths about philosophy in general. A very entrenched myth is the following:

Philosophy is neither right nor wrong, so why bother wasting our time with it?

This is probably the most common myth about the nature of philosophy, and though it may be true for some areas of philosophy (I'm not naming names ...), there are many positions once held in the philosophy of science that are *unanimously agreed* (amongst philosophers of science) to be wrong – we will come across many of these, for their problems are still instructive. For example, Karl Popper's famous position called "falsificationism" – according to which science works by deducing testable consequences from theories (or *conjectures*) and then attempting to *refute* these consequences with experiments, the theory then surviving or dying depending on what happens – is just plain wrong as a "descriptive" account of how science and scientists actually work. For the most part, scientists just don't operate in this way. As a "prescriptive" account (namely, as an account of what scientists *ought* to be doing) it is false too, and perhaps dangerous, for if scientists were to follow this scientific method to the letter, many advances in science would have been lost (we cover this in later chapters).

Next myth (primarily espoused by scientists who refuse to engage with philosophy):

Philosophy is not rigorous and so is too wishy-washy to take seriously!

Again, there are writings of philosophers for which this is probably true (ok, now I'm naming names: Derrida!), but much of philosophy (certainly Western, *analytic* philosophy) is very much rigorous (hence "analytic"). Much of this has to do with the fact that logic, argument, and reason are central to analytic philosophy. Philosophy of science is especially rigorous in this respect, and often goes further by using probability and other branches of mathematics to formalize its arguments – it is in many ways *more* rigorous than much of science. Specializing to philosophy of *particular* sciences (philosophy of physics, biology, etc.) increases this level, and indeed most philosophers of these particular scientific subjects (and of philosophy of science in general) generally have a background in some science or other, as mentioned in the preface.

Last myth:

> Philosophy is useless, so philosophy of science must be too!

This is a pretty common view too. In order to properly convince you that it is wrongheaded, you'll have to read on, and continue your studies beyond this, and then see what you think afterwards. For now, let me use an *ipse dixit* argument (translation: "he himself said it" – appeal to an impressive/smart person's credibility!). Basically, during every fundamental revolution – in physics at any rate – the scientists involved have considered themselves "natural philosophers": Newton, Leibniz, Mach, Boltzmann, Poincaré, Heisenberg, Schrödinger, Einstein (of course), and many others have written philosophical texts. Ronald Fisher, the statistician who first used randomization as an experimental tool in biology (agriculture, in fact), was also very interested in philosophical issues, especially those having to do with causation, explanation, and laws. Fisher was essentially responding to the work of J. S. Mill, another philosopher who dabbled in many sciences – these ideas were then applied in medicine by Austin Bradford Hill (resulting in the randomized controlled trial), who again was intensely interested in philosophical aspects of causation, evidence, and inference. In each case they themselves directly acknowledge the *utility* of their philosophical reflections in leading them to explore new territory. Indeed, philosophical argument seems to have been *vital* in many such cases. Einstein acknowledges that his special theory of relativity owed much to his reading of David Hume (a Scottish philosopher we will encounter often throughout this book). This shift to philosophy is happening again in physics, since seemingly a new revolution is needed to merge a pair of theories (quantum theory and general relativity) that make apparently very different claims about the nature of space and time – this has resulted in increased dialogue between physicists and philosophers. If you don't understand this, don't worry: the point is, many of the greatest scientists who ever lived have been philosophers as much as scientists, and often the mark of a brilliant scientist is a dual philosophical mindset. If the past is any guide, doing some philosophy will make you a success, and wealthy beyond your wildest dreams! (Well, perhaps that last one is an exaggeration, though many major advances in economics have also had their roots in a philosophical analysis of the foundations of economic theory.)

A First Look at Philosophy of Science

With that little defensive stroke played, let's turn to the more positive matter of actually saying what philosophy of science is. Well, in fact, let's start by saying what it *isn't*.

- It isn't a study of the *history* of science in the sense of looking at how scientists actually made their

discoveries, what the conditions were like at the time, how scientists operated in some period, how their methods (of experiment, of reasoning, of disseminating work, etc.) have changed over the centuries. History of science *does* have a very significant role to play in the philosophy of science – and there is some controversy over just how important it is – but, however they might overlap, they are not the same thing.
- It isn't the *sociology* of science in the sense of a study of the way scientists interact, what kinds of social networks they have, how they resolve differences of opinion on various issues, how they decide which theory to choose to work on, generating consensus when there are many possibilities, and so on. These are interesting and valuable tasks, and again they have a role to play in certain philosophical issues (though the extent is, again, controversial), but it is not the same thing as philosophy of science.
- It isn't the *psychology* of science in the sense of a study of how scientists think, how they mentally reach their conclusions, what goes on in their heads when they create theories, etc. Again, we may use such information to inform our philosophizing about science, but the two subjects cannot be identified.

What distinguishes philosophy of science from these other, certainly very worthy, enterprises? Well, in each of the above cases there are *facts* which are discovered. They follow an empirical method, whether empirically observing the scientists themselves ("up-close and personal," with a notebook to hand), or by looking through texts and other sources, such as notebooks and letters. This is not part of philosophy of science. These are branches of history or of science itself. As a general rule of thumb, we might say: if you have to get up out of your chair to do it then it's not philosophy! This is only a rule of thumb because many philosophers do "get their hands dirty" doing practical stuff too, but this is generally incidental. So, what *is* philosophy of science then?

First, what is philosophy? This is a big question to answer in a little section of a little chapter, but we can approximate an answer by saying that philosophy constitutes an inquiry into the world at the most *general* level possible – see the further readings for good introductory texts. This involves abstract categories such as truth, matter, space, time, causation, mind, morality, reason, etc. But philosophy often focuses in on a particular subject of inquiry, so that we have "the philosophy of *X*" (where *X* = "science," "art," "mind," "biology," "law" – i.e. some subject of inquiry which really can be anything at all – there's even a book on the philosophy of Buffy the Vampire Slayer!). When we focus up close in this way, the "philosophy" aspect signals that we have gone "second-order" ("meta-") in the sense that no longer are we investigating the subject matter of the subject of inquiry. Rather, the subject of inquiry itself becomes a subject of inquiry.

Let's give a simple example. Music consists of various activities – composing, analyzing scores, performing, etc. – but *philosophy of* music looks at these activities and their results (compositions, theories about music, performances, etc.) from a philosophical point of view: it asks what music *is*; whether and how music *represents* the world (or some mental entity); how there can be multiple, *different* instances of one and the *same* piece of music, whether music can truly be "expressive," and that kind of thing. Likewise, the philosophy of science puts science itself, and its products (theories), in the spotlight. It looks at the methods used by scientists to see if they are as reliable as scientists think. It examines fundamental, central concepts (i.e. concepts that are used but not analyzed by scientists) to see if they are justified and how they can be understood – this process is often involved in the process of revolutionary science. It asks what theories themselves are, what they say about the world, and whether they support a unique worldview, and so on.

There are several core parts of philosophy, and we will be mainly concerned with two of these: epistemology and ontology. The former is concerned with issues such as whether scientific theories are true and whether we should believe what the scientists tell us (and so looks at questions of *justification* and *reliability*). The latter is concerned with what the world is like if what the scientists say is true (i.e. if the theories are true). There is also logic and ethics, or "value theory." Ethics is the investigation into right and wrong, and morality and proper conduct, and though it certainly does have a place in the philosophy of science (there is a huge related sub-field called "bioethics," for example), we will not devote any time to it in this book since such issues are generally taught as a subject in their own right. Logic, however, will crop up fairly often in this book, since that is the study of arguments and reasoning. Logic is often used to back up arguments concerning epistemological and metaphysical (and scientific) claims.

Let's give a simple example (suppressing for the moment definitions and complications dealt with in later chapters) of what examining science through the lens of philosophy of science might look like. We consider a problem in how theories are tested and justified. A standard story has it that a scientist develops some hypothesis or theory, and deduces some consequences from it that can be subject to testing. She will then test this hypothesis with an experiment. She will, if she is a good honest scientist, repeat the experiment a number of times, in different conditions (to eliminate any "confounding" influences and isolate the relevant feature) and see if the results are more or less the same, checking if they match the consequence she deduced – more honesty still would involve checking for any bias being introduced by herself. If a (finite) run of tests fits her theory, then she will be satisfied, and will be satisfied that more tests *would* satisfy the theory if she *were* to continue making them: there is something "law-like" going on.

This scenario has a number of assumptions implicit in it: it assumes a certain way of testing a theory, namely by *deducing* observational consequences from a theory and comparing them with the world. More importantly, it assumes that nature will continue to behave in the same way in the future as it did in the past. But what does that mean? In what respects will it behave in the same way? Not *all* respects: different rooms, different equipment, different times, different climatic conditions, etc. But there is, we suppose, some structure that is robust enough to persist despite these differences. The idea that nature possesses such universal regularities or invariances involves the notion of *laws of nature*. Laws of nature will crop up quite often in this book: for example, scientific theories, according to one prominent view, *must* contain laws (indeed *are* laws), as must genuinely scientific explanations. Laws go beyond the available data applying even to experiments not yet performed. So we have a problem: finite data or experiments but infinite scope of laws. Data and experiments cannot be used to *deduce* (i.e. infer with absolute certainty) theories. This is the foundation of the "problem of induction" or "Hume's problem": what justification is there for our theories given that we have only knowledge of the past? How do we know our theories will work in the future? How do we know that what happened in 100 experiments will happen in the next? It is perfectly conceivable that the pattern encoded in a law will alter in the future. This is an old and difficult problem that we consider in the next chapter. The question then becomes: what are these laws? That is also a difficult question to answer, but we also will look at some possibilities, again in the next chapter.

Returning to the first assumption, what justifies the claim that, because some observational consequence was deduced and discovered in a few or more tests, the theory is correct? The consequences might just as well have been derived from some other theory, so the results would be as much a confirmation of this other theory (of which there may be infinitely many possible ones). Rarely are the phenomena observed uniquely capturable by one single theory – sometimes this is believed to be the case, in examples of so-called "crucial experiments," but often some further analysis will show this to be incorrect – this is known as the *underdetermination* problem: the evidence does not determine a single theory, but is compatible with many such.

This is probably easiest to see in the case of theories of highly complex systems, such as the theory of anthropogenic climate change. Given the complexity of the system here (with very high numbers of interacting relevant variables), it is hard to isolate the causal factors to test theories and identify the observed phenomenon (e.g. the undeniable increase in global temperature: figure 1.1) as a consequence of one theory only. There are, of course, all sorts of alternative theories that attempt to capture the same observational phenomenon (the temperature rise) from a different theoretical basis (e.g. cycles theory, pointing to relatively recent Ice Ages and so on, as a result of changes in the Earth's orbit relative to the Sun). This does not mean the anthropogenic theory is not true, of course, and there are various additional pieces of evidence, providing plausible mechanisms, that come into play to capture the accelerating rate of temperature increase; but it is far harder to prove than situations in which we can take hold of the individual variables and "wiggle" them as it were – if true, however, policies to reduce global emissions would then see a concomitant reduction or slowing of temperatures, though quantitative estimates of how much and how long this will take are, once again, complicated by complexity.

This is in marked contrast to the famous puerperal fever (i.e. uterine infection following childbirth) investigations of Ignaz Semmelweis, based on a difference

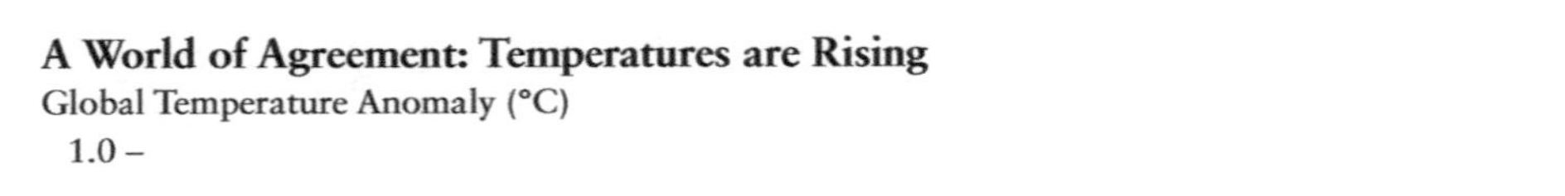

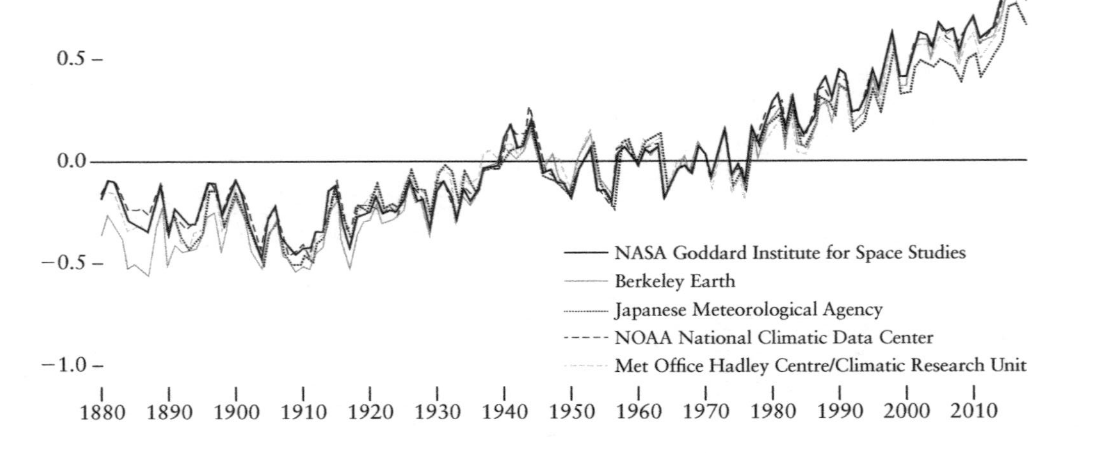

Figure 1.1 Graph showing convergence of evidence that global temperatures are rising, and more rapidly in recent years: the anthropogenic theory would have this effect as an observational consequence

Source: NASA's Earth Observatory/Robert Simmon

Die Aetiologie, der Begriff

und

die Prophylaxis

des

Kindbettfiebers.

Von

Ignaz Philipp Semmelweis,

Dr. der Medicin und Chirurgie, Magister der Geburtshilfe, o. ö. Professor der theoretischen und practischen Geburtshilfe an der kön. ung. Universität zu Pest etc. etc.

Pest, Wien und Leipzig.

C. A. Hartleben's Verlags-Expedition.

1861.

Figure 1.2 The original published presentation of Ignaz Semmelweis's theory of the causes of childbed fever from 1861 (title in English: *The Aetiology, Concept, and Prophylaxis of Childbed Fever*)

Source: Wikimedia Commons

in mortality rates between deliveries performed by surgeons and by midwives in the same hospital (figure 1.2). Here the hypothesis was that an increase in mortality rates amongst new mothers in a maternity ward (whatever was causing it was labeled "childbed fever") was due to the unclean hands of the surgeons, who were also dealing with dead bodies (thus exposing the mothers to "cadaverous particles": a nicer way of saying "bits of dead people") – this, along with the positioning of the mothers (on their side or back) during childbirth, was the only persistently observed difference in conditions that might provide a mechanism grounding such a difference in rates. This gives a clear pair of variables to experimentally intervene in. Make two groups: wash the hands in one, and change the birthing position in the other, and then see what happens to the mortality rates, *making sure that as little else changes as possible.* This was done, and the rates dropped dramatically in the "wash hands" scenario. The control of the variables here allows for the identification of a mechanism of action so that the theory (cadaverous particles somehow cause higher mortality rates) and the outcome (the higher rates) can easily be matched. Of course, this leaves a finer grained description of exactly how the non-washing of hands does what it does, but it clearly isolates a causal pathway.

These are the kinds of issue philosophers of science regularly deal with and the ones scientists usually do not, and probably *should not*, deal with too often – however, the treatment of causality and the making of causal inferences is one area of genuine overlap, and I mentioned that Ronald Fisher (father of randomization as a scientific testing tool) was thinking about just such philosophical issues of causation in his research. The reason scientists shouldn't dabble in these issues lightly is that they are the tools of the scientist's trade: the foundations. Once you start to question them and poke around down in the foundations, the structure becomes unstable (though it might re-stabilize into a different framework). Down this path, it is all too easy to leave science behind and get bogged down in details of how it is that science works at all – a comparison might be with comedians who start questioning why they are funny and how they get laughs: this can be the end of them! Aristotle and Freud analyzed comedy, and they never once got a laugh. Or perhaps it's more akin to how you can walk just fine if you don't think about it, but suddenly lose all ability when you consider what's actually going on.

But, *some* consideration of the art of science is very useful, and if you are an Einstein (or a Semmelweis or Fisher), you can delve extremely deeply into issues such as measurement, causality, and such like (which might be dangerous for some), and get an entirely new fundamental theory or decisive tool out of it! For non-Einsteins, it is still useful to develop some critical skill to weed out problems with current scientific theories, and to diagnose where problems come from, and how they might be patched. For example, you can probe a particular explanation given in some research article and pull it apart, seeing how it hangs together, seeing what assumptions are going into it, seeing if laws are being used, seeing if probabilities are being used (and if they are being used properly). You can probe how evidence is being used, how it was gathered, whether the methods were reliable, whether it is sufficiently strong to allow inferences to some one theory, and so on. You don't need to be a scientist for these kinds of skills to be useful. They can be applied in everyday life, e.g. in having a better grip on news reports about climate, the artificial intelligence "singularity," or the risks of a new drug.

This forms one class of issues, concerning evidence, testing, causality, and laws: epistemological issues. Another major class of issues (ontological in this case: concerning what exists in the world), with less overlap between philosophers and scientists, springs from the fact that science trades in many things that we never directly perceive: genes, market forces, fields, electrons, the Big Bang, curved spacetime (the recent Laser Interferometer Gravitational-Wave Observatory (LIGO) experiments to detect gravitational waves involve direct observations of *mirrors* and light interference patterns, *not* "ripples in spacetime": figure 1.3), unconscious drives, and so on. Do these things really exist, or are they perhaps convenient fictions? Scientists also must refer to things that *can* be observed directly – e.g. things like phenotypic traits, volt-meters, springs, computer

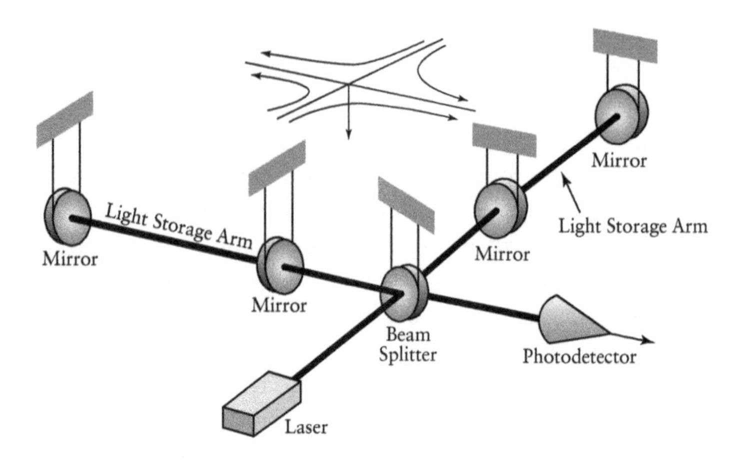

Figure 1.3 A diagram of the observable workings of the LIGO experiment to detect gravitational waves, often phrased in terms of detecting "ripples of spacetime"
Source: Wikimedia Commons

screens, and the LIGO mirrors and optical phenomena – for these are the things ultimately used to test theories and, as is usually assumed, to support belief in the other entities postulated by theories. We have two types of thing here: observable entities and unobservable entities. Philosophers of science will ask whether we are justified to believe in the unobservable entities, and ask for the *grounds* of belief. (We can, of course, ask what grounds we have for believing in the observable entities too, but this is a more general philosophical question.)

What is observation anyway? What does it involve? Do you (a trained biochemist, let's say) see the same thing that I (an utter moron when it comes to microscopes and experimental stuff) see when we both look through an identically prepared microscope? In other words, are our observations (is what we see) conditioned by our background of beliefs, by our training, and by theory? If we can't answer this question, then the division into "observable" and "unobservable" entities seems to be without foundation, yet at least one major position in philosophy of science depends on being able to draw a firm line. Even the "observable" things I mentioned above might be considered "too theoretical," since a volt-meter involves some degree of theory, making it the case that the piece of metal moving is saying something about voltage. Ultimately all we *see* is a dial moving. This kind of problem led the logical positivists to seek the absolute ground-level of observations, which would be as primitive as possible to avoid any theoretical contamination. This they boiled down to such things as "green patch here" and "black line and point coincide there" (these statements would be called "protocol sentences" or "atomic statements"). The unobservable statements were supposed to be grounded in these atomic propositions, from which they received their ultimate justification in an extreme form of empiricist foundationalism.

There are a host of related issues too: science involves idealization, abstraction, massive simplifications in order to make problems more tractable – frictionless planes, perfect harmonic oscillators, etc. This is a feature of most, if not all, sciences. Why are we justified in using these? The bizarre thing here is that these idealized models are known to be *false* (e.g. we know perfectly well there *is* friction in our world), and yet we use them to "discover" the world all the same. How can this be? The laws of science contain such idealizations too, so strictly speaking, as Nancy Cartwright so nicely puts it, they *lie* about the world – see her book *How the Laws of Physics Lie* (Oxford University Press, 1983).

It's fairly obvious that philosophers are going to have an interest in science then: philosophers are interested in knowledge and truth. Scientists often claim to have a reliable method for generating knowledge and truth. Much of philosophy of science is devoted to the assessment of these claims. For instance, in chapter 3, we consider whether there is a solid way of distinguishing the claims of science from non-science (especially mysticism, religion, etc.). Surely, if science has this special status, it should have some way of so distinguishing itself, right? It turns out there are many possibilities, but none is entirely satisfactory. This is just a small sample of topics which barely scratches the surface of philosophical investigation into science and the sciences. But hopefully you can begin to see something of the nature of the problems that philosophers of science deal with, and why they are of importance for our understanding of the world. And, for that matter, why scientific understanding might not be quite as transparent as is often supposed.

What Science Cannot Do

There are certain "metaphysical" issues bequeathed to philosophers that seem further out of the reach of the sciences themselves. For example, mathematics deals with numbers of various types, and patterns relating them, and so on. But though mathematicians can compute many wonderful things, they cannot tell us what a number is, and what the patterns are patterns *of*. Are numbers *things*? Are they objects like tables and chairs? Are they simply constructions of the mind? Mathematics can get on perfectly well without answers to these questions (just as physicists can get on without asking whether there are *really* ripples in spacetime), but if we are interested in why mathematics works, what makes certain mathematical statements true, why it applies so well to our world when theories are formulated mathematically, and so on, we turn to philosophy. In this sense, philosophy is a "deeper" discipline.

Another example, which introduces more obviously philosophical issues, comes from physics. Newton's first law of (classical) mechanics states that force is equal to the product of mass and acceleration: $F = ma$. Yet acceleration is "temporally loaded," since it itself is equal to the time rate of change of velocity. But then we face the question: what is time? Also, what is mass? This is not so easy, and can refer to an intrinsic property of an object, or a resistance to a force (inertia), or whatever it is that couples through gravity. Other quantities in Newton's laws introduce similar questions to do with space, matter, and causality. Newton was well aware of these philosophical issues, and was led (vicariously through his supporter, Reverend Samuel Clarke) into a philosophical debate about the real nature of space and time with Leibniz: a debate that still has consequences for how we formulate present-day theories of physics (including the as yet non-existent theory of quantum gravity!). The nub of the debate is whether or not space and time are entities with a mode of existence separate from material objects. Interestingly, there have been many attempts to dislodge space, time (or spacetime), and matter from philosophy, into the realm of physics, but so far they have resisted in many respects, though philosophers have had to take on board the developments of physics. (The problem is, the theories they come up with are riddled with metaphysical problems, often to do with identity: for example, in what sense are spacetime points, included in all spacetime theories, *things* at all given that they can share *exactly* the same (intrinsic) properties: surely differences require distinctions?)

Hence, science brackets certain fundamental questions in order to get on with the business of solving things and understanding things at a lower level of abstraction. Philosophy deals with the questions that science cannot answer without stepping outside itself to observe it. There is another aspect to this: in order to succeed in science these days, scientists have to specialize to the extreme: they most often occupy a tiny bit of a fragment of some field. This restriction means that they cannot cover the whole of their "mother-subject." If they can't even cover this, or sometimes even a subfield of this, then they certainly cannot venture outside of their mother-subject. To do so is often frowned upon in fact, though, as mentioned, some major names have done so. But this state of affairs means that the scientists themselves cannot view the whole of science, they cannot see how the various sciences hang together. This is an important issue that does need looking at, and philosophers of science have become the ones who do it, without risking their careers in the process! In being more generalist, philosophy can see the forest, while sacrificing some of the details of the trees.

We might ask, though, can (or should) science answer all questions? We might think that if a question cannot be answered by science, then it is simply not well-posed: a pseudo-question. But are questions such as "what is causation?" or "what is a number?" really pseudo-questions? If they are, then that is a view that needs defending (as indeed it was, by the logical positivist school). Any attempt to defend it will result in the introduction of philosophy, and metaphysics at that. For the claim that "science can answer anything that isn't a pseudo-question" requires a well-founded distinction between "genuine questions" and "pseudo-questions." This is a philosophical venture: one cannot merely say that whatever cannot be answered by science is thereby a pseudo-question! That begs the question of what pseudo-questions are, and to state that they are those that cannot be answered by science is blatantly circular. Even if we could set up this distinction, there is still a normative issue lurking: why should science be dealing with the non-pseudo-questions? This is a philosophical question: there is really no escape from philosophy! Though note that this isn't restricted to professional philosophers: scientists can and do answer these questions too, but in so doing they step into philosophers' shoes.

Making Sense of "The Sciences"

Philosophy and science have had an interesting relationship, rather like an initially perfect marriage gone bad, leading to their inevitable divorce – we might extend this unhappy analogy by viewing philosophy of science as the offspring, not entirely sure of which parent its allegiance should be aligned with. The history of science, from the Babylonians to the ancient Greeks onwards, has involved the progressive detachment and independent establishment (as separate domains of inquiry) of various subject matters and methods once the sole domain of philosophy. Philosophy really used to encompass pretty much *every* domain of inquiry one could think of! It was concerned with knowledge in all its forms (hence *Φιλοσοφια* = *philosophia* = "love of wisdom"). However, in the third century, mathematics (initially via geometry thanks to Euclid) detached and became "the science of patterns" (or *space* if we restrict ourselves to geometry). Likewise, in the seventeenth century, physics emerged as a discipline largely independent of metaphysics (though it is often referred to as "natural philosophy" – Newton, for example, viewed himself as a natural philosopher, though mathematics played the central role). Since then many other disciplines that seem even closer to philosophy have split: cosmology, psychology, cognition, decision theory, logic, and even studies of morality and ethics. What's left for poor old philosophy? Hopefully, I've said enough already to make a case for philosophy of science. However, let's say something more, by focusing on what science cannot do, but philosophy can.

To say that something is "science," or that something has been carried out "scientifically," is generally understood as bestowing a great honor on that something. This even infects advertising of "scientifically proven" shampoos and such like. This is perhaps due to the fact that science is distinguished from other fields in that it is based on methods (the "scientific method") that are supposed to lead reliably to the truth (or some approximation thereof). It gets its results in a way that is supposedly independent of human frailties, bias, imprecision, ambiguity, and so on: in this respect it is *objective*: you can trust it. Where elements of imprecision, error, uncertainty, and bias do creep in, they are dealt with scientifically too: that is to say, objectively and precisely and without bias.

Indeed, a famous skincare company landed itself in trouble over claims that its products were based on scientific principles. L'Oréal claimed that its Lancôme night care cream "boosts the activity of genes," in fact doubling such activity (whatever that means!) (see figure 1.4). The US Federal Trade Commission investigated such claims, finding them unwarranted, subsequently doling out fines for attempting to push its products under the auspices of scientific research. While L'Oréal's claim "Because I'm worth it" was in the clear, because of difficulty in proving otherwise (being a purely subjective value judgement), stating that gene activity doubles on application of a cream is a testable one – and is, in fact, nonsense.

Science is often presented as the most *rational* of disciplines. In this it is contrasted with, for example, religious devotion which involves faith, and supernatural studies which also seem to involve "leaps of faith." Science is supposed to be "special" from an epistemic point of view in that it leads us reliably to truths about the world – we will see, especially in our discussion of scientific realism in chapter 4, how there are problems with this view; we also consider a "patch" for these problems. There are some – mostly sociologists of science (who observe the behaviors of scientists, rather than assessing the veracity of what they produce) – who hold that science *isn't* special from an epistemic point of view: it is just one amongst many human systems of belief

Figure 1.4 Skincare cream leading to a lawsuit over the unlawful use of unwarranted scientific claims

Source: https://www.ftc.gov/news-events/press-releases/2014/06/loreal-settles-ftc-charges-alleging-deceptive-advertising-anti

(like the voodoo of the Haitians or the Shamanic rituals of certain South American tribes). In case you like the sound of this view, I will try to nudge you a little away from it in later parts: the point is that modern science is just too predictively successful and self-consciously controls its errors and biases too well to be so lightly compared to these other "systems of belief." But there are problems with the notion of *the* scientific method: chief amongst these being that most of the great discoveries (of Newton, Darwin, and Einstein) follow no such simple and similar method. It must also be admitted that while scientific reasoning is indeed highly distinct from, for example, voodoo, it comes from a similar source in the human mind, namely a desire to make sense of the world. Historically, too, science grew out of systems of thought infused with magic, mysticism, and the occult.

We will bracket these issues for now, and consider the more basic question: what is distinctive about science? What distinguishes it from what we might call "pseudosciences"? We answer this question in two parts, beginning in this section with an overview of science, and then focus in on the philosophical issue. So, before we launch into a philosophical investigation of science, we should first get to grips with what it is (though this is, in effect, already a philosophical question). (We will see how this becomes a *legal* issue when we get to an important religion-based case study in chapter 3.) Here are two possible answers to the question "what is science?":

- Science is whatever scientists do.
- Science is anything carried out according to a "scientific method."

There are obvious problems with the first option: scientists do *many* things! What are the ones we are interested in? We might say that what we mean is that science is whatever it is that scientists do *that is common to all and only scientists*. This is intended to get to the core of science. But the problem now is that there are many *types* of science, and it is unlikely that there is a common core that could reasonably cover all of these types. Moreover, what then is a scientist? Surely this must be defined in terms of some pre-existing notion of science, and so we have simply gone in a circle. Perhaps the second answer comes closer: the common core is the *methods* used. But now the question is: what are these methods? There are two problems here: (1) there are a great many methods throughout the sciences, varying a great deal depending on the particular science; (2) what *distinguishes* these methods from other methods in the non-sciences?

An old and still popular view is that science follows the "inductive method." The core of this view is that science is based on experience: this is known as *empiricism*, and is to be contrasted with *rationalism* (the view that true knowledge of the world comes from pure reason). We start with observations. We can write these as "observation statements":

- The electron had charge *e*.
- The litmus paper turned red when it was put in the liquid.
- Your leg looks bent when it is in the swimming pool.

These are gathered from experience. What is important about them is that they are *singular statements*: statements referring to particular events (particular electrons, litmus, and legs) at specific moments of time and specific places. Science (the natural sciences at least) does not deal in such weak statements. It uses **laws**: universal statements referring to *all* instances, at *all* times and *all* places in the universe!

- Electrons have charge *e*.
- Acid turns litmus paper red.
- When light rays pass from one medium to another,

they change direction such that the sine of the angle of incidence divided by the sine of the angle of refraction is a constant characteristic of the pair of media.

The view involves several contentious assumptions. One we have already met, that observation provides a faithful record of what is happening in the world (it isn't clouded in any way). More seriously perhaps, it involves a massive leap from a finite number of observations to a general law that applies universally! That is a problem: how do we justify this leap? This forms the basis of Hume's problem, or the problem of induction: how do we get from experience, from the singular observation statements, to the universal statements of science? This is a killer problem: our theories, predictions, and explanations are made up of laws, universal statements, so how are they justified? If we can't answer this, then science, if we follow inductivism and empiricism, is in trouble.

The inductivists do have an answer that involves placing a number of conditions on inductive generalization:

- large numbers of observations,
- testing in different conditions,
- no conflicting observations.

But this still isn't sufficient: the inductive generalization is *universal* (i.e. infinite in generality). Science, if inductive, cannot account for its own success!

Another part of science is that it involves explanation and prediction. There is a close relationship between these concepts (as we see in the next chapter). So, given our theory (replete with laws), we can explain or predict by combining the laws with some observation statements (giving us our initial conditions describing the way the world is at some time). We can then *deduce* (yes, no longer inductive but deductive) some consequence. If the consequence has never before been seen, then we have a prediction. If it is a known phenomenon we are trying to capture, then we have an explanation: *X* happened because we have this theory from which it can be deduced. The theory enables us to derive the phenomena, so it is explanatory: it grounds the phenomenon. So, the model is that we use induction to get at the laws and theories, by making lots of observations and doing experiments. From the laws and theories (and some initial conditions) we deduce some consequences. If the consequences are then observed we have a prediction (if it is novel) and an explanation. Hence, explanations are simply deductive arguments on this view. Take an example: Newton's universal law of gravitation. We might make lots of measurements to see how gravity affects bodies. We derive a law and theory from these: $F_{grav} = G\frac{m_1 m_2}{r^2}$ (i.e. the gravitational force between two bodies, 1 and 2, is equal to the gravitational constant *G*, describing the strength with which bodies couple, times the product of the bodies' masses divided by the square of the distance that separates them). Once we have this law, we can deduce consequences; we can make *predictions* and test the theory. So for *these* two bodies (a pair of planets, say) with masses m_1 and m_2 positioned a distance *r* apart (where these values we input are the initial conditions), we can see if they satisfy the motions that we would expect given the truth of the law. If they do, the prediction is confirmed, and so is the theory. The theory then explains why the planets move the way they do. You probably think this sounds good: it does. But there are serious problems with it as we will see in the next chapter, and as you might already be able to figure out from the preceding discussion.

Some Prehistory and History

Though our subject is philosophy of science, we learn much about it from an examination of its life story:

where did it come from? How did it get to look the way it now does? Why do we focus on the kinds of question that now occupy us? The story is more exciting than you might think, and intersects with revolutions in science and in politics, and even in art – of course, we barely scratch the surface here, but see the further readings section for good accounts of this history.

The origins of modern philosophy of science hark back to the 1920s, to a bid to produce a thoroughly *exact* account of knowledge of the world, by the so-called "Vienna Circle." The model for exact thinking was the mathematical sciences, and it was suggested that other forms of knowledge creation should follow the same kinds of approach. What couldn't be manipulated to fit this model would have to be discarded as nonsense or mere pseudoscience – so much the worse for religion, psychoanalysis, and metaphysics. While this approach, known as "logical positivism" (or "logical empiricism"), is now widely considered to be completely dead, its legacy lives on in very many ways in contemporary philosophy of science, and indeed much of what we discuss comes from this old approach, if only as responses to its problems.

However, the roots of our subject go back further than this, to the birth of modern science itself in fact, and the Vienna Circle's own line of thinking emerged from a rich background of earlier work – not least our old friend David Hume's. And, there are far earlier examples of something like philosophy of science independent of this. We briefly review this "prehistory" before tracing the story, with a very coarse-grained periodization, from the Vienna Circle to our current era, which, despite certainly coming a long way, as mentioned still bears many of the hallmarks of these interesting origins.

The birth of modern science is generally located roughly in the two-century-long period between the middle of the 1500s and the middle of the 1700s, and is usually labeled "the scientific revolution," though some contemporary historians of science do not like this term and often can be found denying the existence of anything of the sort – while I tend to agree with them, we'll put such matters aside for a simpler life (see the further readings for more information). What was considered to be revolutionary was the specific *method* employed in discovering new knowledge, and a shift away from the older non-experimental Aristotelian methods. But this brought also revolutions in world-view (i.e. the physical picture we have of the universe). Perhaps the key event triggering this so-called revolution was Nicolaus Copernicus' (1473–1543) discovery that the Earth did not lie at the center of the universe, but instead orbited a central Sun (figure 1.5) – in fact,

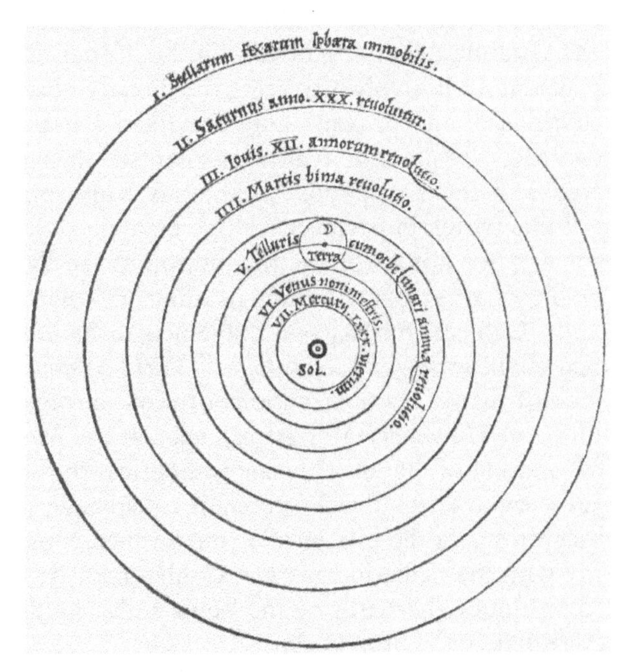

Figure 1.5 Copernicus' diagram of the structure of the universe from De Revolutionibus Orbium Coelestium ("On the Revolutions of the Celestial Orbs"), published 1543
Source: Wikimedia Commons

Aristarchus had a heliocentric model over 1000 years earlier. Since Aristotelian science was grounded in the geocentric model (later formalized in a precise model by the Roman astronomer Ptolemy), in denying this, Copernicus started a revolution, not least because it had religious ramifications, dethroning the Earth. Though Copernicus' book was dedicated to the Pope, it was banned by The Vatican not long after publication, along with other books claiming that the Earth was not perfectly stationary at the center of the universe as per the scriptures. Much subsequent development of science amounted largely to explorations and extensions of this Copernican model.

There *is* something truly radical about Copernicus' idea (despite the preface added by his publisher, Andreas Osiander, claiming otherwise, for Copernicus' own protection): it involves treating the directly observable motions of the heavens (and the fixed Earth) as "appearance," while the reality is that the heavens are fixed and the Earth moves! Thomas Kuhn made much of this switch in perspectives (see below). Making sense of why we don't observe the motion of the Earth was Galileo's great contribution (leading to what we now call "Galilean relativity"), explained in his book *Dialogue Concerning the Two Chief World Systems* – well, one of many: his observations of celestial objects (recounted in *The Starry Messenger*), such as the Moon and the moons of Jupiter, revealed imperfections and change, in contrast to the Aristotelian worldview; and his observations of the phases of Venus seemed consistent only with the Copernican view. Galileo really made it hard to doubt the truth of the heliocentric models, using writings of the utmost clarity.

Subsequent developments involved the progressive mathematization of theories of the world, itself championed by Galileo who famously claimed, in *The Assayer*, that "the Book of Nature was written in the language of mathematics." According to Galileo, this mathematical representation of the universe was necessary in order to make it comprehensible by the human mind: it is how we *understand* the Great Book of Nature. The culmination of this approach was the invention of calculus as a modeling tool, invented by Leibniz and Newton around the same time – the name "calculus" comes from Leibniz. However, it was Newton's mathematization efforts, especially in his *magnum opus*, the *Principia Mathematica*, that are often said to have established "modern science" as we know it today, and indeed to constitute another revolution all of its own. Indeed, the Newtonian worldview was considered so well-established and certain that Immanuel Kant considered the mathematical structures for space, time, and causality involved in Newton's theory to be the way we *must* view the world to have experience of it (forming what Kant called "the categories"). Though today's theories are certainly more complicated mathematically than Newton's theories, we largely follow the same kind of approach – though, for better or worse, with far less focus on supplying an underlying metaphysical picture than Newton: despite having a predictively successful theory of gravity, Newton didn't have a metaphysical basis grounding gravity, which he believed was required to have an actual explanation. Of course, the twentieth-century revolutions of quantum mechanics (overthrowing the old theories of matter, energy, and force) and Einstein's theories of relativity (overthrowing the old theories of electromagnetism, space, time, and gravity) would radically modify the Newtonian worldview. In large part, these new theories left a vacuum for a new approach to philosophy of science and scientific method in their wake. This was filled by the views of the scientific philosophers known as the logical positivists, but before we get to them, let us first shift our historical focus to the notion of a scientific method.

Fields of inquiry rarely spring forth into the world without some precedent. We can find scattered throughout the historical record all kinds of tracts on *method* in science, or in reasoning about the natural world. In less scrupulous works on the history of science one might find the claim that Francis Bacon "invented" the scientific method. We have to be careful with such claims right away, since the very terms at stake ("science" and "scientific methodology") have shifted their meanings over time, if they existed at all in earlier times. The most sensible thing to say here is that, strictly speaking, there is no such thing as *the* scientific method and, inasmuch as it exists at all, it is something that shifts with the times. Even in the ancient myths of the Babylonians and Sumerians, we can find something like primitive versions of the kinds of reasoning we associate with the scientific method. Take, for example, the curious fact of the existence of many languages in the world. That was a puzzle to the ancients: it is indeed a puzzle. The myth of the Tower of Babel (the tower of *confusion*) is supposed to provide an explanation, however bizarre, as follows: following the Great Flood, the surviving people spoke one language, and working together set to building a tower tall enough to reach heaven. God confounds their plan, by introducing a "Babel of languages" and scattering these people across the Earth. Whatever you might think of this, it does offer an explanation: the story's conclusion is identical to the puzzling phenomenon, a multiplicity of languages, and has a certain logic of its own.

However, the term "method" indicates something more rigorous than merely coming up with stories to explain the phenomena, that is, with mere *accounting for*: there must be *rules* involved, more like an algorithm than an allegory. Francis Bacon is often viewed as "the father" of the modern scientific method in this sense. There is some substance to this. His treatise entitled (or rather subtitled) *Novum Organum Scientiarum* (the "new instrument": figure 1.6) was a direct attack on the old Aristotelian system of reasoning about the world

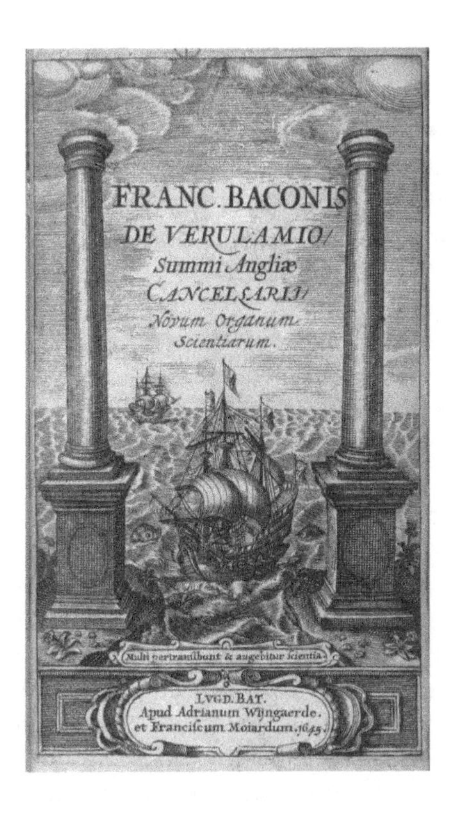

Figure 1.6 Title page for *Novum Organum Scientiarum*, 1645, by Francis Bacon (1561–1626). EC.B1328.620ib, Houghton Library, Harvard University: Francis Bacon of Verulam / High Chancellor of England / New Organon

(a hybrid empirical/deductive logical system which his disciples had named the "organon"). This new organ of Bacon's was intended exactly as an instrument or machine that one employs to generate knowledge about the world in a reliable manner. A primitive version of something like an experimental method, with an attempt to isolate proper causes and effects and so on, was part and parcel of this method. This certainly contributed heavily to the surge of new work that occurred during the period known as the Scientific Revolution. The aim was precisely to go beyond "mere observation" by attempting to control for the biases of the human mind and other accidental conditions of the observations (which Bacon called "Idols") that might confound the observations.

You might find it curious that it was ever doubted that one finds out about the world by poking it and seeing what happens. But, while Aristotle was himself in fact largely an empiricist, believing that knowledge of the world must come via the senses, what was done with Aristotle's views departed from this foundation, especially when they were purloined by the Church. However, the difference between the "organs" really has to do with the kinds of *logic* they employ. For Aristotle, the deductive syllogism (taking us from general statements to particular ones) is at the core of proper reasoning about things; for Bacon, it is *induction* (taking us from particular observations to general truths) that leads us to knowledge of the world. While Aristotle had a role for induction, it was in generating some broad principles that would then be involved in the deduction of certain other facts about the world, so that the final move to knowledge was deductive. Moreover, unlike the Baconian method, there was no control involved in the inductive generation of facts, but merely passive observation (inasmuch as such a thing is possible, in light of what was said above).

Despite this Baconian impact, Isaac Newton developed his own very stringent scientific methodology (explicitly laid out in his "Rules for Philosophising" in Book 3 of his *Principia Mathematica*), going from effects to causes *by deduction*, but involving the inductive method too in examining the causes (the appearances or phenomena) – in this, he was arguing in direct opposition to Descartes' anti-empiricist (rationalist) methods. A nice explanation of his method can be found in his *Optiks*:

> The main business of natural philosophy is to argue from phenomena without feigning hypotheses [pure guesses

> about the nature of the unobservable – DR], and to deduce causes from effects, till we come to the very first cause, which certainly is not mechanical; and not only to unfold the mechanism of the world, but chiefly to resolve these and such like questions (Newton, *Optiks*, p. 369).

In his researches on light, Newton utilized his methods to great effect, inducing that white light is made up of a spectrum of colors (each with its own angle of refraction) by using a prism to split sunlight – crucially, this involved further testing his inductive inference that light is *composed* of these colors, by attempting to repeat the splitting with a second prism on the individual colors from the first prism, finding that they do not in fact further split and so are indivisible components of white light (and not, e.g., some artefact of using prisms). This secondary stage (which Newton called "the method of synthesis," a deductive stage following the initial inductive stage or "method of analysis") is closely related to what would become the "verification principle" of the logical positivists: for any claim to have meaning, it must be testable with experiment or observation; though the positivists do not require any constraints on how a hypothesis or theory is initially generated. However, it is clear that Newton was not consistent in his methods, and his first law of motion (that speaks of "bodies on which there are impressed no forces") cannot have been the product of induction, since no such bodies exist (e.g. *all* bodies observed have at least a gravitational force acting upon them).

The eighteenth-century philosophers David Hume and Immanuel Kant are other important ancestors of modern philosophy of science. We have already mentioned Hume, and will return to him later. Kant divided the world into "knowable" (phenomenal) and "unknowable" (noumenal) realms. While we can know the *appearances* of things (the way the world appears *to us*), there is a deeper layer, the "true" nature of things as they are in themselves independent of us, that we can never know. In the nineteenth century, Auguste Comte applied Kant's idea to scientific theorizing. He thought that if we accept Kant's view, then the empirical sciences (based on observations) cannot lead us to the truth. Nor *should* they. This is not what science is about. It is just about observations, and predictions about what new observations we will make in the future. We should not think about what happens in the murky shadows behind these observations. This stance Comte labeled "positivism" – similar statements were made around the same time by several others. At the same time as these positivistic, anti-metaphysical ideas were floating around, so were new developments in logic taking form, especially by Bertrand Russell and Gottlob Frege. These suggested the possibility of exact thinking and a foundation for mathematics. Given the mathematical nature of much of science, perhaps logic could provide secure foundations for scientific knowledge too? Indeed, until just before the middle of the twentieth century, there was a widespread belief that philosophy should deal with science only from the logical perspective, and indeed that philosophy was "the logic of science" (that is, the logical analysis of concepts, propositions, proofs, theories of science, etc.). Logic then offered an ideal core to a new approach to science that outlawed metaphysics in favor of a "positivistic theory of knowledge."

Such was the outcome of meetings between a remarkable group of thinkers in Vienna in the 1920s – from this exuberant time came modernist architecture, atonal music, abstract art, Wittgenstein, and much more. The real establishment of a "school" of philosophy occurred in 1928, with the production of a manifesto outlining the principles of the new theory of scientific knowledge. At the core of this approach was the positivistic thesis, known as the principle of verification, stating that the meaning of a statement is given by its method of verification. If this is not forthcoming, then the statement is rendered meaningless. This is a kind of reductionism amounting to the claim that unless you can turn a statement into a statement about some possible experience, that statement is meaningless. A very simple refutation of logical positivism, due to Carl Hempel (whose work we meet in the next chapter), is that it violates its own verification principle, for it tells us that any statement that cannot be verified by the methods of science is meaningless. Doesn't this make the approach self-refuting; hoist by its own petard? This is rather a cheap shot. But there were eventually other serious, internal problems that emerged (not least problems having to do with a split between the empirical or synthetic statements of science and the analytic statements of logic and mathematics), and certainly the original members moved away from a strict adherence to this principle.

As the pre-war tensions in Europe caused the dispersal of the Vienna Circle across the globe, an influential philosophy of science journal (cleverly entitled *Philosophy of Science*) was founded in America in 1934 by William Malisoff, with several émigré Vienna Circle members involved. (In fact, the Vienna Circle established its own journal, *Erkenntnis*, which was ended as a result of the war; a journal with this title was started up again in 1975, but with a rather different outlook.) This journal set, and continues to set, much of the agenda for philosophy of science, and within its pages logical positivism morphed into logical empiricism, and was progressively developed and destroyed, leading to what we find today: a less logic-focused subject – of course there were other relevant journals involved too.

Karl Popper was always slightly outside the Vienna Circle, despite being a native of Vienna himself. His own non-inductive approach, similar to logical positivism in providing a kind of criterion of meaning, denied induction the special role it had inherited from the days of the Scientific Revolution. Indeed, Popper's key objection to logical positivism stemmed from its inductive nature, which led to logical problems. Instead, he viewed the progress of science as a rather heroic venture in which scientists come up with hypotheses (which are never verified by experience) and then valiantly attempt to slay them with experiments: as a matter of logic, you can't be sure you have the right theory, but you can be sure you have the *wrong* theory.

Although scientists like this image of how they operate, it does not quite respect the historical record. A series of works starting in the 1960s offered a new way of doing philosophy of science, with more focus on the *realities* of scientific practice and history – this started an ongoing battle between the more logical approaches and historical approaches, which wasn't really broken until relatively recently. One of the central differences of these newer historical approaches is the greater focus on the so-called "context of discovery" versus the "context of justification." On the logical approaches, of both the positivists and Popper, the way science was produced was acknowledged to likely be a messy affair, not following the logical principles they laid out, but the way the resulting product was assessed ought to obey firm logical principles. Psychology could deal with the precise details leading to the discovery itself: it might come from a dream (Mendeleev and Kekulé); a psychedelic trip (allegedly, Francis Crick); a vision from God, or who knows what. None of this matters. How it is tested is all that matters. This mindset is now considered to be rather old-fashioned.

A key architect of this new approach was Thomas Kuhn, primarily in his landmark book *The Structure of Scientific Revolutions* from 1962. Kuhn's approach caused trouble for almost every tenet of the previous approaches stemming from the logical positivist tradition, and also the idea of such a thing as the scientific method itself. The basic idea of Kuhn's approach to science is that there are distinct *phases* to the development of science which have very different qualities: most

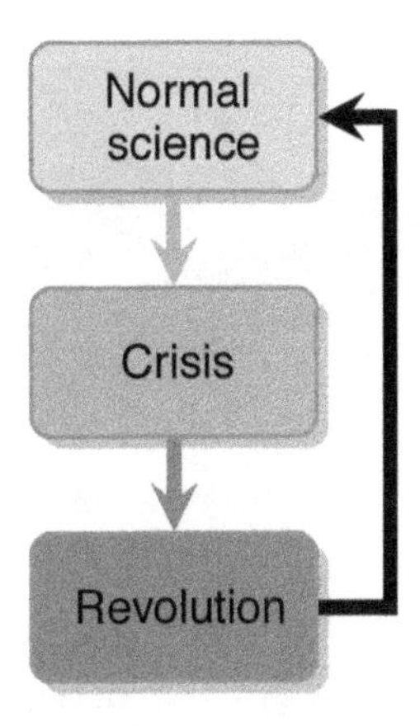

Figure 1.7 Normal Science, Crisis, Revolution

science proceeds inductively, with little questioning and a steady accumulation of knowledge. Problems can build in this stage, until eventually a crisis emerges. This requires a serious rethinking about fundamentals (one might say this is a philosophical stage). A revolution occurs that then establishes a new pattern for normal science to carry on as before, although now in a new *paradigm*. This can be represented by the flow chart shown in figure 1.7.

A key idea is that after each cycle, when a new order or paradigm is established, the new normal science is "incommensurable" with the old: they offer genuinely different worldviews that disallow comparison. If taken seriously, this has massive ramifications across the standard issues in philosophy of science, not only with respect to scientific change (which now has the form of a punctuated equilibrium), but also realism, since the worldviews are changing too. In other words, science is not now a case of a progressive accumulation of knowledge, but can be obliterated and replaced with something new – so new, that the previous ideas fail to make sense from the new vantage point. This is often described as a kind of "gestalt switch," much as one

Figure 1.8 Joseph Jastrow's ambiguous "duck-rabbit" image, used by Thomas Kuhn as a kind of metaphor for paradigm change in scientific revolutions
Source: Harper's Weekly (November 19, 1892, p. 1114)

finds in the famous "duck-rabbit" picture (see figure 1.8) or the Necker cube.

It is obvious that science changes. One of the mantras that public scientists, such as Richard Dawkins, proffer is that science is never 100% foolproof: it is fallible, and proudly wears this fallibility on its sleeve – in stark contrast to the claims of the faithful, for example. Sometimes the changes are slight, like an adjustment to the value for the charge of an electron. Other times they are more profound, like the switch from understanding heat as a fluid (in caloric theory) to heat as an aspect of the motion of particles (in the kinetic theory of gases). This poses very serious problems for any claim that science leads us to the objective truth about the nature of the universe, for how do we know that we ever have the correct picture? And if these pictures are indeed radically different, we cannot even speak of getting closer and closer to the "true picture." These questions triggered much recent work, still ongoing, and were pivotal in a new way of thinking about scientific theories in terms of *models* of reality that represent in various ways, and that contain all kinds of idealizations and simplifications. Of course, this brings new problems of its own.

This very brief excursion has missed out a great many details, and has likely simplified in part almost to the point of inaccuracy. However, the following chapters will delve in greater detail into many of the topics raised, and draw out many subtleties missing in this chapter's rough and ready treatment. We start in the next chapter by looking at the role of logic in the philosophy of science, and at how it has been employed (and subsequently criticized) in the context of four core areas: (1) induction and inference; (2) confirmation and evidence; (3) laws of nature; and (4) explanation. This core group of topics then infiltrates the remaining two chapters, on the problem of demarcating science from other disciplines and on the nature of scientific theories and how they relate to the world.

Summary of Key Points of Chapter 1

- Philosophy is not a case of "anything goes," and philosophy of science has a great many virtues: it provides rigorous methods to scrutinize the scientific enterprise using the tools and methods of philosophy.
- Two important areas of philosophy involved in philosophy of science are "epistemology" (the study of the grounds of knowledge of the world) and "ontology" (the study of what exists in the world).
- The orthodox "inductive" way of thinking about the scientific method (taking us from observations to theories), commonly ascribed to Francis Bacon, is fraught with complications, not least the fact that theories outstrip observations. We might not be justified in believing what scientists tell us about the nature of reality.
- Modern philosophy of science harks back to an intellectual movement (logical positivism) devoted to making philosophy as close to a scientific subject as possible, drawing from logic and natural science,

with the core principle that to be meaningful is to be verifiable by observation. Recent philosophy of science has been in large part a reaction to logical positivism.

Further Readings

Here are some excellent little books introducing readers to philosophy. Together, these books would quickly bring beginners up to speed on how philosophy works, and how to start doing it properly:

- Timothy Williamson, *Doing Philosophy: From Common Curiosity to Logical Reasoning* (Oxford University Press, 2018).
- Simon Blackburn, *Think: A Compelling Introduction to Philosophy* (Oxford University Press, 2005).
- Bertrand Russell's classic little book: *The Problems of Philosophy* (Oxford University Press, 2001) – this is still a lovely reading experience.

On the linkages between science and philosophy, see:

- Lucie Laplane, Paolo Mantovani, Ralph Adolphs, Hasok Chang, Alberto Mantovani, Margaret McFall-Ngai, Carlo Rovelli, Elliott Sober, and Thomas Pradeu, "Opinion: Why Science Needs Philosophy." *Proceedings of the National Academy of Science* (2019) **116**(10): 3948–52.
- Carlo Rovelli's call for philosophers can be found in "Physics Needs Philosophy / Philosophy Needs Physics." *Scientific American Blog*: https://blogs.scientificamerican.com/observations/physics-needs-philosophy-philosophy-needs-physics/.
- A superb debate on The Role of Philosophy in Science from the *Moving Naturalism Forward* workshop, October 2012 (with participants including Sean

Carroll, Jerry Coyne, Richard Dawkins, Terrence Deacon, Simon DeDeo, Daniel Dennett, Owen Flanagan, Rebecca Goldstein, Janna Levin, David Poeppel, Massimo Pigliucci, Nicholas Pritzker, Alex Rosenberg, Don Ross, and Steven Weinberg) can be found at https://www.youtube.com/watch?v=zZny-Zqcok4.

The history of philosophy of science has been quite widely studied, especially as it concerns the Vienna Circle. Some of these texts are quite advanced reading, and reveal many controversies in making sense of what happened and who influenced whom. But others are more general.

- A fairly detailed examination of the emergence and early development of modern philosophy of science is Friedrich Stadler's "History of the Philosophy of Science. From Wissenschaftslogik (Logic of Science) to Philosophy of Science: Europe and America, 1930–1960." In T. Kuipers (ed.), *General Philosophy of Science: Focal Issues* (Elsevier, 2007), pp. 576–658.
- A great collection of classic texts on logical positivism can be found in A. J. Ayer (ed.), *Logical Positivism* (The Free Press, 1959).
- Viktor Kraft, *The Vienna Circle, the Origin of Neo-positivism; A Chapter in the History of Recent Philosophy* (Greenwood, 1953).
- More general historical treatments of the philosophy of science and scientific method in particular are Barry Gower's *Scientific Method: An Historical and Philosophical Introduction* (Routledge, 1997); John Losee's *A Historical Introduction to the Philosophy of Science* (Oxford University Press, 2001); and Richard DeWitt's *Worldviews: An Introduction to the History and Philosophy of Science* (Wiley-Blackwell, 2018).
- An impressive historical treatment of philosophy of science, using modern text mining techniques to

reveal historical trends, is Christophe Malaterre, Jean-Francois Chartier, and Davide Pulizzotto's "What Is This Thing Called Philosophy of Science? A Computational Topic-Modeling Perspective 1934–2015." *HOPOS* (2019) **9**(2).

- For a more general history of science, with a philosophical orientation, a modern classic on the scientific revolution (or the absence thereof) is Steven Shapin's *The Scientific Revolution* (University of Chicago Press, 1996).
- For an excellent, though rather advanced treatment of the mathematization of physics from a historical point of view, see Yves Gingras' "What Did Mathematics Do to Physics?" *History of Science* (2001) **39**(4): 383–416.
- In terms of online resources (of which there are many of high quality, but vastly more of not so high quality), a good place to start is to look through the entries in *The Stanford Encyclopedia of Philosophy*: plato.stanford.edu. I will refer to specific entries in subsequent further readings sections. I will also be referring to videos that are on YouTube at the time of writing, choosing only videos that are unlikely to be removed, though this cannot be guaranteed. A good start here is to watch the short videos on "Why Philosophy of Science?," from the PBS TV series *Closer to Truth*: closertotruth.com/series/why-philosophy-science-part-1. See also Bryan Magee's discussion of philosophy of science with Hilary Putnam, from his superb 1978 BBC TV series on *Men of Ideas*: youtube.com/watch?v=h7Z2y61rd6M.

2

Logic and Philosophy of Science

Modern day philosophy of science is, for better or worse, impossible without some grasp of logic. Many of its key ingredients are couched in logic. For this we have our old friends the logical positivists (or logical empiricists) to thank. This in fact provides us with a very nice, unified account of a whole bunch of key concepts which view various facets of science in terms of logical relationships. To a large extent, this kind of treatment pushed the more non-mathematical sciences to the margins for several decades, with physics taking pride of place. The demise of logical positivism brought with it a greater focus on sciences other than physics, especially biology which is now, within the philosophy of science, perhaps even more dominant than physics.

Making Inferences

A common mythical picture of scientists has them deducing theories about the world from the gathering of facts. This is the way of Sherlock Holmes, who was created precisely to be a "scientific detective" by Arthur

Figure 2.1 Picture from the Sherlock Holmes story "The Adventure of the Naval Treaty," by Arthur Conan Doyle (illustration by Sidney Paget). The image caption reads: "Holmes was working hard over a chemical investigation."
Source: Wikimedia Commons

Conan Doyle. A famous image has Holmes performing chemical tasks, with Watson looking on (figure 2.1). It is perfectly true that several of the practical methods used by Holmes made their way into forensics (e.g. fingerprinting and footprint analysis), but the methods he employed were not *deductive* as is claimed. If anything, they were what philosophers would call "abductive," or "inferences to the best explanation." There is no *unique* logical link between Holmes' theories and the evidence. For example, in *Silver Blaze* there appears the famous "curious incident of the dog in the night-time": despite what should have been a noisy event of the horse being led out of the stable block, the nearby dog did nothing – a null fact that leads Holmes to his suspicions that the horse was not stolen by a stranger, for the dog *should* surely have barked. This is, then, at best an inductive inference rather than a deduction – though, at the time of Doyle's writing, "deduction" referred to inference more generally. It is based on facts about dogs and what they *usually* do. This style of reasoning is nicely encapsulated in Holmes' dictum: "when you have excluded the impossible, whatever remains, however improbable, must be the truth" (eliminative induction).

Much of science takes a similar form of inferences from evidence to theory. However, the approach of the Vienna Circle and its descendants *did* involve deductive logical relations (in the modern sense), though between statements of fact, rather than evidence and theory. Note also that often scientific discoveries are made in a way seemingly at odds with this Holmesian approach, and can involve guesswork and all manner of intuitive approaches to come up with a theory. The logical positivists would then detach this theory from its context of discovery and focus their attention only on how it was justified. Holmes' method, and indeed other inductive methods, assume that hypotheses themselves must also be subject to the rigors of the scientific method, just as much as their justification.

Scientists tell us lots of things that go against common sense, that we would not otherwise believe. They tell us that we are related to apes; that the universe is expanding; that there is no single Now separating past, present, and future; that the continents used to be locked together in a giant super-continent known as Pangaea (see figure 2.2). Why do we believe them? How did the scientists

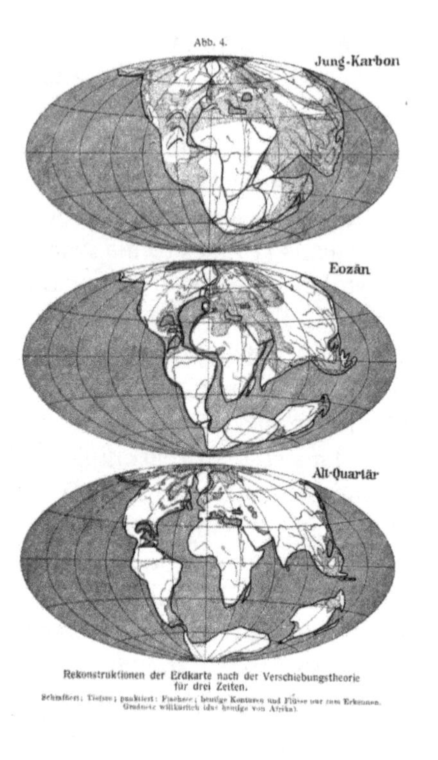

Figure 2.2 Alfred Wegener's reconstruction of the supercontinent of Pangaea as based on his theory of plate tectonics: *Die Entstehung der Kontinente und Ozeane* (The origin of continents and oceans), 1929, 4th edn

Source: Wikimedia Commons

themselves come to these conclusions? After all, these are not the kinds of thing one can directly observe. They do it, of course, by employing a method: the fabled "scientific method." They arrive at their beliefs by a process of reasoning or *inference*, much as with Holmes, though with greater attention to the reliability and bias-free nature of the inferences. But what exactly is this process, and why is such confidence placed in it? In this chapter we answer these questions and show that, according to David Hume, our confidence in such an inferential picture is badly misplaced!

Problems of Induction

According to empiricists, all of our information comes from observation (*Nihil in intellectu nisi prius in sensu* – "nothing in the understanding that did not get in there through the senses"). Inductivism is based on observation: this is the foundation-stone of inductive approaches. According to what we might call "naive inductivism," science starts with observation, this observation provides a *secure* base on which scientific knowledge is supported, and scientific knowledge is derived from observation using induction (inductive inferences). Rationalists, on the other hand, are not wedded to observation: some knowledge (about the world) can come from pure reason alone. So, according to empiricists, our knowledge is *justified* by our experience (observation, data, experiment). The objectivity and rationality of science is taken to rest on the role experience plays in choosing between hypotheses and in justifying those hypotheses.

Empirical observations, in the context of science, are explained by hypotheses of a general kind: the hypotheses apply to *all* of a class of events or phenomena, only a sample of which have or will ever be observed. Given this, how can we be sure that some theory that performs this explanatory function is the right one? There are surely many such possible theories that would do the job as well.

This is the problem of induction: how do we get from empirical observations to scientific theories? The Cambridge philosopher C. D. Broad called induction "the glory of science and the scandal of philosophy." We will see why that is still the case. Before we get to it, we have some initial material to review. (Note that this is by far the longest chapter in this book, since it contains most of the core issues that forged philosophy of science into what it is today. They are treated as a unity since the problems come from the same source, in the specific logical setup of central scientific concepts.)

Some Words on Logic

Since the problems we are going to deal with have a distinctly logical aspect (though the problem is really epistemological), let's now say something about logic and, especially, the difference between deduction and induction. Logic, in a nutshell, is about good and bad reasoning. Since ordinary language is often imprecise, it is difficult to assess reasoning in terms of it: so we (us philosophers) have to resort to formalism or, sometimes, just supplying very precise meanings to certain words. Here we simply review some of the more basic elements.

Firstly, what is an argument? This is one of the most basic notions in (scientific) reasoning. An argument consists of a set of premises (one or more) and a conclusion. The idea is that the premises give *reasons* for the conclusion. The premises are propositions: they can be either true or false. Good arguments are those such that the premises give good reasons for the conclusion and the premises are known to be true. Things go wrong, and we have a bad case of reasoning, when the premises do not support the conclusion.

Serious errors in logic are called "fallacies." A classic example is *affirming the consequent*. We will see this in action in the next chapter, when we look at ways of demarcating science and pseudoscience. One response to this latter problem is that science follows a "method." In particular, a popular account says it follows an *inductive* method: gather data and generalize from the data to make general laws; if an instance is found that backs the law, then we have confirmed the law. This is wrong, since the instance could have occurred in many ways, regardless of the law. An everyday example: "If it is raining, then the road will be wet," "the road is wet," therefore "it is raining." This is false: the road could be wet for any number of reasons (a hose-pipe, a water fight, a broken fire hydrant, etc.). Here "it is raining" is known as the "antecedent" and "the road will be wet" is known as the "consequent." If we'd have said: "If it is raining, then the road will be wet," "it is raining," therefore "the road is wet," then that would have been a good argument (it has its own fancy name: *modus ponens*): if the premises are true, then the conclusion *has* to be true.

Though the argument is rock-solid in terms of validity (i.e. the conclusion must be true if the premises are – this is a logician's term; it should not be confused with the ordinary-language usage of "valid"), we might still question the premises: and this would be a way of pulling the argument apart. Most of philosophy consists of either demonstrating invalidity of arguments or, if that fails, showing one or more premises to be false. Note that logic, however, is only bothered about reasoning, about the link between premises and conclusion; it doesn't care so much about truth and falsity. Propositions can be true or false; arguments are valid or invalid. Only deductive arguments are valid. An inductive argument is invalid as a logical argument; even though its premises might well be true (and so might its conclusion), they are not sufficient to allow us to infer the truth of the conclusion, as we will see.

Let's give some simple examples of inductive and deductive arguments. Remember, a deductive argument is just one such that the truth of premises implies the truth of the conclusion – and in this sense, the premises "contain" the conclusion already. An inductive argument is simply one for which this isn't the case. An example of a deductive argument (called, in this case, a syllogism) (where the line separating the statements stands for "therefore") is:

All Yorkshiremen drink real ale
Dave is a Yorkshireman

Dave drinks real ale

This is a perfectly valid deductive argument: if it is the case that All Yorkshiremen do indeed drink real ale (ale in a cask, for those not in the know), and, furthermore, it is true that Dave is a Yorkshireman, then it must be true (as a matter of logic) that Dave drinks real ale. Obviously, this does not make the argument true (or *sound* in logicians' terminology). Validity and soundness (truth or falsity) are two completely different matters. So, if someone were to use this argument to try to really argue that Dave, some Yorkshire person whose personal drinking habits he doesn't know, drinks real ale, we can simply say that one of the premises is not true: all Yorkshiremen do *not* drink real ale! The argument, though valid, is therefore unsound. The meaning of validity here can be understood along the lines that taking the premises to be true and the conclusion to be false leads to an inconsistency: it is a contradiction to say the premises are true and the conclusion false – this is the meaning of a deductively valid argument: true premises must take you to a true conclusion. In other words, if All Yorkshiremen *really do* drink real ale, and if Dave really is a Yorkshireman, then it must be the case that he drinks real ale *as a matter of logic*!

Now let's consider an *inductive* argument:

All the Yorkshiremen I've ever met drink real ale
Dave, who I've not met, is a Yorkshireman

Dave drinks real ale

We might actually use this argument if we were planning to stock the fridge for a Yorkshire visitor. And then the chain of reasoning above would apply. But

Figure 2.3 A Timmy Taylor's beer tasting session in Yorkshire (1937). While all the Yorkshiremen here do indeed drink real ale, we cannot generalize from this finite sample to all Yorkshiremen. In this case, we cannot infer that Dave, a Yorkshireman whose drinking habits are unknown, drinks real ale

Source: Timothy Taylor & Co. Ltd, reproduced with their kind permission

the argument is quite manifestly invalid. We can hold the premises to be true and yet deny the conclusion without contradicting ourselves: there is no inconsistency involved. Dave could be the exception to the rule, and might in fact be partial to a Campari and soda. I might only have met five Yorkshire folk (see figure 2.3), for example, and that is no solid basis to generalize on. A valid generalization would involve meeting all actual and all possible, past, present, and future Yorkshire chaps and knowing their drinking habits! (In fact, in many sciences, *laws of nature* are invoked to perform exactly this function: covering all possible scenarios, and so enabling generalizations to unseen cases; but we face a similar problem of justifying them in the first place – we return to this below.)

So, when we reason deductively, we can be certain that if our premises are true then the conclusion must be true. Not so with inductive reasoning. But deductive reasoning doesn't really apply to interesting real-world situations. For this we use induction. Yet induction is not capable of taking us from true premises to a true conclusion. But we rely on it on a day-to-day basis: we couldn't really get by without it. When you wake up in the morning and step out of the bed to get up, you rely on the fact that the floor is going to be there. Why? Because it has been there on every other occasion in the past and floors seem to be the kinds of entity with persistence and stability (unlike sand dunes, say). When you go to walk out of the door to leave your house, you expect the rest of the world to be out there. Why? Because it hasn't failed to be there yet! These are both inductive inferences: we reason from a past, finite number of actual cases (and plausibility, itself based on similar considerations), to future cases. But these are cases of logically invalid reasoning! Science, too, is based on this kind of invalid reasoning (at least according to a "standard" account – Karl Popper, as mentioned, thought that it was, or at least *should be*, all deductive). Why is this the case? Well, just think about what laws are: they are statements that concern *all* events of a specific kind. This involves an infinity of cases. We can never have observed an infinity of cases, so the inference must be from a finite number of cases to an infinite number of cases: this is inductive. So scientific claims are, in general, never *proven*, they might have evidence in their favor, but certainty is not usually possible (unless you follow Popper's view that certainty can be achieved by denying a decisive role for induction in science). Let's now consider one of the classic philosophical problems concerning inductive reasoning.

The Problem of Induction

C. D. Broad, in his paper "The Philosophy of Francis Bacon," wrote that:

> There is a skeleton in the cupboard of Inductive Logic, which Bacon never suspected and Hume first exposed to view. Kant conducted the most elaborate funeral in history, and called Heaven and Earth and the Noumena under the Earth to witness that the skeleton was finally disposed of. But when the dust of the funeral procession had subsided and the last strains of the Transcendental Organ had died away, the coffin was found to be empty and the skeleton still in its old place. Mill discreetly closed the door of the cupboard, and with infinite tact turned the conversation into more cheerful channels. Mr Johnson and Lord Keynes may fairly be said to have reduced the skeleton to the dimensions of a mere skull. But that obstinate *caput mortuum* still awaits the undertaker who will give it Christian burial. May we venture to hope that when Bacon's next centenary is celebrated the great work which he set going will be completed; and that Inductive Reasoning, which has long been the glory of Science, will have ceased to be the scandal of philosophy? (C. D. Broad, from An Address Delivered at Cambridge on the Occasion of the Bacon Tercentenary, October 5, 1926).

The philosophical problem of induction questions the very possibility of inductive reasoning. It began with David Hume. Hume published his *Treatise of Human Nature* in 1739, at the age of 28. It is one of the classic works of all time (I'm a bit of a fan myself), and in it we find the problem of induction, also now called "Hume's Problem" as a mark of respect.

Hume was a skeptic. In ordinary English, a skeptic is just someone who constantly doubts accepted beliefs (religious beliefs, for example), or just mistrusts people in general. Hume was a *philosophical skeptic*, however, which goes farther than this. A philosophical skeptic is one who doubts any real knowledge or sound belief about X (where X is something of philosophical interest: a field of knowledge or a belief of a moral, scientific, or religious nature). Skepticism of this sort is particularly strong: it means that *nothing* could ever help with the problem! Hume was a skeptic about induction, and his argument is supposed to provide reasons for this.

Well, firstly, we should note that *everyone* uses induction all the time, not just in science but in everyday goings on, as mentioned above. So to say that induction can never be justified is looking a bit serious! Although inductive reasoning is not absolutely certain reasoning, surely it's a rational and sensible way to reason? Indeed, if we met anyone who thought the floor might not be there when they stepped out of bed in the morning, or that the world would not be there when they stepped out of the door, we would most likely think they were mad – we would not think they were hyper-rational. Where does our faith in induction come from, given that it is a deductively invalid form of reasoning? Hume basically said that there can be no rational justification of induction. The person who reasons that the world might not be outside of the house is *more* rational in a certain logical sense – however, Hume explicitly advocated acting like other people in the common affairs of life and not being pulled into the philosophical delirium for too long, instead allowing nature to take over and playing a game of backgammon or conversing with friends to dispel the clouds that come from this overly rational stance!

Inductive reasoning can be applied to past, present, and future. We can ask why the dodo became extinct, and do some historical research on that. We can ask what somebody is doing in the other room right now, and reason about that. Or we can ask whether someone will attend a lecture next week. Generally, when people talk about induction, they mean this "future directed" notion. However, in general, induction involves making inferences from a particular case, or set of cases, to a general claim. From a finite sample of data about something, to a claim involving all of a class of things. This might involve an inference from a class of past things to some future things. This happens a lot: we infer things about the future, based on present knowledge, because we (think we) know that the future is like the past (i.e. the future has been like the past *in the past*). This is based on a principle Hume calls "the uniformity of nature": roughly, unexamined things will behave like similar examined things.

A first pass at Hume's problem is: How do we know the future will be like the past? Or, How do we know the principle of the uniformity of nature is actually true? We might think that's easy: the future is like the past because present things *cause* future things. But Hume's skepticism applies to causes too! Hume asks: Where do we get this idea of causation? He gives an example involving billiard balls (figure 2.4). Imagine a billiard ball lying on the table, and another one moving towards it with some rapidity. The balls strike, and the ball that was at rest now moves. This, says Hume, is a perfect example of the cause–effect relationship (as perfect as any we know through "sensation" or "reflection" – Hume's terms for knowledge gained through the senses and reasoning). No problems so far. Now, the two balls touched one another before the motion was passed on,

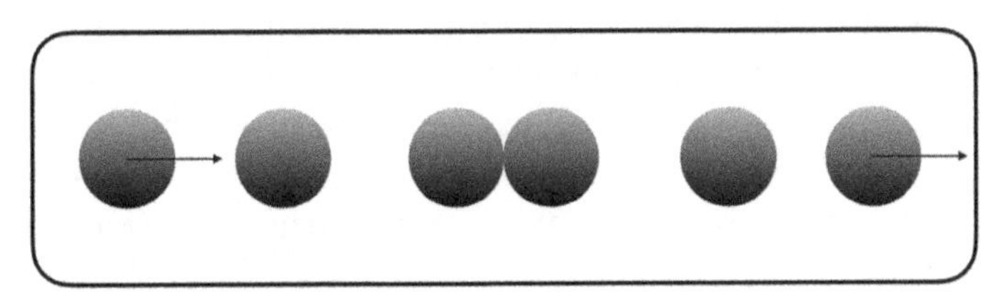

Figure 2.4 The three stages of a billiard ball interaction according to Hume: (1) the first ball approaches the other, (2) the balls touch, and then (3) the second ball moves away from the first

and there was no interval of time between the striking and the motion (otherwise they would both be standing still, so why would the other ball then move?). We can say that the motion that we take to be the cause of the other ball's moving was before that other ball's moving (the effect). That is a condition of causation: causes precede effects. Also, if we were to run this little setup again (with different balls), we would find the same chain of events: like causes produce like effects. Hume calls this "constant conjunction." We have three aspects here: touching ("contiguity"), priority (of cause before effect), and constant conjunction. "Beyond these three circumstances" says Hume "I can discover nothing in this cause." We just have: one ball moved towards another non-moving ball; they touched; and then the initially non-moving ball moved, with the earlier moving ball now stationary. Whenever we try this out, with similar balls in similar circumstances, the same thing happens. That is *all* there is to the idea of cause and effect from an empirical point of view.

Hume argued that reasoning about cause and effect is based in experience, on experiences of like following like, of constant conjunction: it is just habit. Inductive reasoning in general is based on the same idea, that "the course of nature will continue uniformly the same." But what this means is that cause and effect cannot justify our belief that *the future will be like the past* because cause and effect are founded on this assumption of uniformity, i.e. the assumption that the future will be like the past! This is just what we're trying to explain with cause and effect. So causation cannot be the answer. More simply, we can note that this principle of uniformity is not necessarily true: worlds which change randomly without warning are perfectly conceivable. So, if we could *prove* the principle of uniformity of nature, then such worlds would be utterly inconceivable (i.e. logically impossible). And, as we said above, if we try to demonstrate the truth of this principle using empirical evidence (i.e. by reasoning inductively), then we would have to assume the very principle we are trying to prove.

Can we use an inductive argument to justify our belief that the future will be like the past? The future has been like the past in the past, so it will continue to be this way in the future! No: this is *viciously* circular; it rests on the same inductive assumption that we're trying to justify.

Trying to defend induction by saying it has done well up till now, so why shouldn't it continue to do so, is to beg the question: it assumes what's at stake. It would only work if you already had a justification for induction. Trying to persuade in this way somebody who didn't believe in induction would not work.

And yet science, according to the "received view," is based on induction. Experience is supposed to give us a firm epistemic basis on which to build our knowledge of the world, from which we then deduce our great theories, and on which basis we build bridges, airplanes, and all that. Science is supposed to be the epitome of rational inquiry, but Hume's problem seems to show this belief to be utterly unfounded.

Bertrand Russell put the problem in the form of an amusing, but rather gory, story. The story involves a turkey who is a firm advocate of the inductive method. The turkey arrived at the farm in summer, and on his first morning found that he was fed at 9 a.m. sharp. But the turkey was an inductivist remember, so one morning isn't enough to generalize from. The turkey was naturally unsure if he would be fed again next morning. But sure enough he was: 9 a.m. again! The turkey, by now a master of the inductive method (a neighboring pig had lent him a copy of Bacon's works), had eventually made a large number of observations under a very wide range of circumstances (different days, rainy days, sunny days, etc.) and always the result was the same: feeding at 9 a.m. sharp. The turkey had a logbook of data showing the conditions and the result of each day:

a massive set of observation statements. The turkey had done enough to make any inductivist proud. Finally, one day (December 24) the turkey felt he'd done enough to make an inductive inference, a generalization from his catalogue of observations to the conclusion that "I will always be fed at 9 a.m." Next morning the turkey's throat was cut and it found itself on the farmer's table as Christmas dinner! This morning he was not fed at 9 a.m. An inductive inference with true premises led the turkey to a false conclusion. What could he have done differently?

Responses to Hume's Problem

Maybe seeing that like follows like lots of times in the past lets us say that the future will *probably* be like the past? The idea is that, although the truth of the premises of an inductive argument cannot prove the truth of the conclusion, they can serve to make the conclusion more probable (i.e. more probable than if the premises were not true).

Hume considers this possibility himself. He argued that probability involves the principle of uniformity too: it implies that the probabilities are going to work the same in the future. Probability has played a key role in many attempted responses to Hume's problem; we consider more below (there's much more to the story than Hume's treatment suggests). Probability does not offer us a unique interpretation, so it is difficult (a serious philosophical problem) to work this out. Part of the problem with probability is that the concept admits multiple interpretations. The *frequency* interpretation, for example, says that when we speak of probabilities, as in the expression "the probability of an Australian woman living to the age of 100 is 0.1," we really just mean that one-tenth of Australian women live to be 100. Likewise, "the probability that a male smoker will get cancer is 0.25" means simply that a quarter of all male smokers develop lung cancer. But this interpretive strategy doesn't always work. Take the following example: "the probability that there is life on Mars is 0.00001." What does that mean on the frequency account? It must mean that one out of every ten thousand Mars has life on it. But that doesn't make sense: there's just one Mars, and only nine planets in the solar system. Frequencies just don't make any sense here, nor with single shot events like the Big Bang or the Cambrian Explosion.

A possible interpretation that can make sense of such probability statements is the *subjective* interpretation. On this account the statement "the probability that there is life on Mars is 0.00001" simply expresses the belief of the person who utters it about the likelihood of life on Mars. I have different degrees of belief (different levels of confidence) in various things: I think it very likely that I will live beyond 60; not so likely that I will live beyond 100; extremely unlikely that I will become immortal by uploading my mind into a computer system; and so on. Here, there are no *objective* matters of fact about probabilities: there's what *you* believe, what *I* believe, and what others believe about the likelihoods of various things. Though there might well be a matter of fact about whether there is life on Mars, that is quite independent of our beliefs about it; there is no matter of fact about how probable it is.

Then there is a *logical* interpretation that says that there is a matter of fact how probable life on Mars is *relative to some body of evidence*. On this account the probability of a statement is a measure of the strength of evidence in its favor, rather than a measure of the strength of someone's confidence in it.

Do these solve the problem of induction? Not all of them would: the subjective interpretation is hopelessly inadequate here. If there is no such thing as objective probability, then it makes no sense to say that conclusions of inductive inferences are objectively probable.

The frequency interpretation requires that we know the frequencies. This requires induction; it requires our saying that a high proportion of some things have been observed to have some properties. So we just have another inductive argument: from all examined instances of some kind being a certain way to it being probable that *all* instances of the same kind are the same way. The logical interpretation might work. For if there is a high objective probability that the world will be outside my door when I open it, then it is surely rational to believe: induction would then be justified by the fact that there is this high objective probability grounded in a body of evidence. It will allow us to say that the person who worries about the floor not being there when he steps out of bed is irrational, and the person who doesn't worry about such things is rational. The former person knows that the evidence that the floor has never failed to be there yet confers a high probability on it being there again. So, here the probability of a statement is a measure of the evidence in its favor. The problem with this is not so much the solution itself, which is a very nice idea; rather, it is the fact that a good account of the logical interpretation cannot be given in the first place – this is still an open problem.

It might be possible to evade Hume's problem of induction. If we can agree that Hume is right, that there is absolutely no justification of inductive inference, but nonetheless claim that it is not important, then we will have evaded the problem. We can do this by showing how we don't need inductive inferences anyway!

One formerly popular response accepts the idea that induction cannot be justified, but says that induction is so fundamental to how we reason and think that it is not the sort of thing that we could ever justify. It is the sort of thing that *conditions the way we think*. This response was defended by Peter Strawson, a famous Oxbridge philosopher and Kant scholar (this response is in his book, *Introduction to Logical Theory* (Methuen,

1952)). He defended this view using the following analogy: suppose one was worried about whether a particular action was legal (Arkansas' Act 590, for example: see the case study in the next chapter). One would simply go to the law-books and compare the action with them. No problem: the law proceeds by such comparisons with precedent. But what if one was worried about whether the *law* itself was legal? This is a weird thing to wonder about because the law is the very thing that one uses to determine whether something is legal or not. The law is the standard against which the legality of things is judged; and it makes no sense to inquire whether the standard of legality is itself legal: against what standard? Ditto induction, says Strawson. Induction is one of the standards we use to assess whether claims about the world are justified. It is too deep in our reasoning to inquire about issues of justification. Basically, acting according to the evidence, and structuring one's beliefs on the basis of evidence, is, in large part, what it *means* to be reasonable. Consider how a meter is defined (or, at least once upon a time): it is just the length of a particular rod in Paris. Can we ask how long the meter itself is? Not really: the meter bar grounds the units of length, and it is ratios between other lengths and it that determine length. Being reasonable is a matter of applying similar standards to arguments to that, and questioning it amounts to doing something improper.

This is a typical piece of "ordinary language" philosophy: look at how a certain concept is used in common discourse, and base the sense on that. Does this response work? Well, yes and no. As we said, this is an *evasion* of the problem rather than a solution: it agrees that the problem cannot be solved directly. So, in this sense, Hume's problem still stands. What the response does is to say that we should stop worrying about it, because it is the kind of thing that could never admit of a justification.

One thing that is problematic is that the proposal seems too vague and general. We can accept that it is indeed reasonable to apportion one's degree of belief to the strength of evidence, but what do we mean by "strength of evidence"? When is evidence strong, when is it weak? Strawson says "quantity" and "quality" are what count: the more evidence there is under more varied circumstances, the stronger the evidence is. This just seems too broad to be of any use. There are plenty of examples one can devise to show that this explanation fails. I might visit a country in summer and see that all the leaves on the tree are green, on many and varied trees. This is looking like strong evidence for the hypothesis that all leaves are green *always*, according to Strawson's account. We need background knowledge to know that this is not the case. Where does this evidence come from if not more evidence? Is that evidence strong? A regress threatens that leads us right back into the jaws of the original problem. Another problem is to do with the fact that saying that, as a matter of definition, "being reasonable" includes in its meaning an acceptance of induction means that this has no empirical content whatsoever, and is purely linguistic.

Popper agreed with Hume that there can be no justification of induction. He, too, had an evasion up his sleeve though. He says we should distinguish between two questions concerning induction: a psychological one and a logical one:

- **Psychological Question:** Why, as a matter of human psychology, do we make inductions? Nothing to do with *human* psychology, says Popper: stupid bacteria do the same thing; they respond to their environment and learn from experience. Like Hume says, this is just custom and habit and there's nothing strange about it.
- **Logical Question:** Don't humans have reasons for our expectations that the stupid bacteria don't have?

> No, say Popper and Hume: there can be no justifiable reasons. Besides: we don't need to make inductions, says Popper.

Popper's idea is that the only good reasoning there is is deductively valid reasoning: we can, moreover, *make do* with this kind of reasoning. So Popper says that the philosophical (or logical) problem of induction is solid; but, never mind, because we don't need induction anyway. Popper naturally also accepted the force of the problem of induction as a problem about theory verification: scientific theories cannot be conclusively verified. So how is it that scientific theories can be superior to their pseudoscientific rivals? Popper's response was to argue that though scientific theories cannot be conclusively verified, they can be refuted (falsified) as a matter of logical form. Also, scientific theories have deductive consequences that can be tested through observation. Testability and falsifiability are paramount. For example, consider the stock-in-trade example of a general statement: "all swans are white." This has certain observable consequences that can be deduced; namely, that the next swan you observe will be white (indeed, for *any* thing, if that thing is a swan then it must also be white). Clearly, observing just one non-white swan will refute the whole claim. And, indeed, in Australia one can find such a swan to *refute* the generalization, thus falsifying the white-swan conjecture (see figure 2.5).

A problem with this (one of the many problems) is that though such refutations may "slim down" the class of possible theories that are compatible with (or "corroborated by") the evidence, an infinite class still remains. We will meet other problems with Popper's proposal in other chapters. However, another point worth considering now is whether, even if this did succeed, it would show that science is rational after all. Surely rationality has to do with success? Unless the success of our theories is covered by rationality, then securing rationality seems

Figure 2.5 Black Swan (*Cygnus atratus*). Illustrated by Elizabeth Gould (1804–1841) for John Gould's (1804–1881) *Handbook to the Birds of Australia* (Lansdowne Press, 1972)

a little empty. Why be impressed by the rationality of science if that rationality has nothing to do with how well theories do? Being impressed by not losing or being able to lose seems peculiar from this standpoint.

Inductive inferences tend to take us from some set of examined or observed instances of something (some event happening or some correlation between some things or properties, for example) to all instances. The form is like the following:

All *x*s seen so far have been *A*s

All *x*s are *A*s

This, along with deduction, is often seen to exhaust the available patterns of reasoning. However, there is another pattern that appears to fit neither of these.

Consider the following inference involving Colonel Peacock and his butler, Jenkins:

(1) Colonel Peacock was found dead in the library with a knife in his back
(2) Jenkins was seen walking about with a knife earlier
(3) Jenkins was seen running from the library with blood on his hands just before Colonel Peacock was found dead

---

The butler did it!

This is a pattern of reasoning known as "inference to the best explanation" (called "abduction" by the philosopher C. S. Peirce, who thought it constituted a genuinely novel pattern of inference, like deduction and induction but distinguished in important ways). Obviously there are lots of possible conclusions that we could stick into the conclusion spot without inconsistency: aliens might have come in and knifed Peacock, then Jenkins came in and found them, the aliens did some kind of mind-control jazz on Jenkins putting blood on his hands and making him run from the room. The Colonel might have fallen on the knife after Jenkins gave it to him for hunting, and the blood on Jenkins' hands might have been from preparing dinner. You get the picture. The point is, this is not deductive reasoning: the premises can be true and the conclusion false. But the hypothesis that Jenkins did it is surely looking *better* than the aliens hypothesis: so the conclusion that Jenkins did it is the best explanation out of a potentially infinite class of possible explanations.

The thought here is maybe we don't need inductive inference because we really use this other form of inference. The inference is obviously non-deductive. However, how does it relate to induction? Does it really help us evade that problem? Maybe, if one could show that (1) this is how we reason, so that we don't need induction; (2) this does not face a similar problem to Hume's problem; and (3) this is not based on induction in some way. (1) and (3) are clearly bound together, if induction to the best explanation is really based on induction, then we can't do without induction after all. Firstly, let's deal with (2): it clearly doesn't face Hume's problem because it is not, on the surface anyway, a generalization from particular to general. Though it does not lead to *certain* conclusions, we don't need it to: we just need the best explanation from a list of possible explanations (cashing out what we mean by "best" here is a problem, and might involve such criteria as "simplicity," which is then also tricky to spell out). Now (3): does this form of reasoning involve induction? On the surface it doesn't: for, as I said, we have no generalization going on. However, a little look beneath the surface reveals that induction is at work: it involves the belief that aliens have not been known to do this kind of thing in the past, while butlers have; it involves the belief that there is a causal connection between the blood on Jenkins' hands and stabbing incidents; and so on. These are all based on inductive reasoning. It seems that this is just good old-fashioned induction by a different name.

This is yet another evasion response – this involves probability once again, but it is given a more sophisticated treatment. We again accept that Hume is correct that given some premises based on experience that are supposed to give reasons for some conclusion, it is possible to form any opinion at all. However, there remain things to be said. The question is, are our opinions, along with various degrees of belief we have about those opinions, *rational*? Are we being reasonable when we modify these degrees of belief when some new information comes in, when we have new experiences, new evidence? The response we consider now – the Bayesian response – says Yes! It invokes a theory of "personal probability," such that the structure of our beliefs satisfies certain (very reasonable, and quite commonsensical) axioms of probability. This allows us to say that there is a uniquely reasonable way to learn from experience (using something called "Bayes' Rule"). So: Hume is right, on this account, but that doesn't matter; all we need is a model of reasonable change in belief. This is enough to guarantee the rationality of our actions in a changing world. Here, "degrees of belief" are represented by numbers between 0 and 1 (0 is absolute uncertainty and 1 is absolute certainty). These degrees of belief need to satisfy basic laws of probability (otherwise they will be incoherent). If they satisfy these laws, then the Bayes' Rule follows, allowing us to "update" earlier degrees of belief in the light of new evidence (experience), and it operates in a coherent and rational way.

Now, suppose that I am interested in some hypothesis *H* and some possible piece of evidence *E*. I also have a prior opinion on how likely *H* is, represented by the probability of *H*, *Pr*(*H*) – this is my prior personal betting rate on *H*. In my set of beliefs there is also something corresponding to my *conditional* betting rate that *E* will occur *given H* (i.e. conditional on *H*'s being true). This is represented by $Pr(E \mid H)$: the probability of getting *E* if *H* is true (or, the likelihood of *E* in the light of *H*). If we really do have degrees of belief like this for the various kinds of possible hypotheses that interest us (scientific ones, for example), then Bayes' Rule tells us that our betting rate *after* learning *H* (represented by the *posterior probability* $Pr(H \mid E)$) should be proportional to the prior probability, $Pr(E \mid H)$, times the likelihood: $Pr(H \mid E)$ is proportional to $Pr(E \mid H) \times Pr(H)$. No less a figure than Henri Poincaré defended something like this position, stating that "the physicist is often in the same position as the gambler who reckons up his chances [where] [e]very time he reasons by induction he more or less consciously requires the calculus of probabilities ... [I]f this calculus be condemned, then the whole edifice of the sciences must also be condemned" (see his wonderful book, *Science and Hypothesis* (Dover, 1905), pp. 183–6).

Hume's own answer was simply that we are creatures of habit: we expect the future to be like the past because of inductive habits we have. "We are determined by *custom* alone to suppose the future conformable [to be like – DR] the past." We might well be able to give an evolutionary explanation for this too: creatures with this habit have done better than those without it! However, justification is still nowhere to be seen: how are the habits justified? Hume says they are not (not by reason): induction is just something we happen to use.

> [T]he experimental reasoning itself, which we possess in common with beasts, and on which the whole conduct of life depends, is nothing but a species of instinct or mechanical power, that acts in us unknown to ourselves; and in its chief operations, is not directed by any such relations or comparisons of ideas, as are the proper objects of our intellectual faculties. Though the instinct is different, yet still it is an instinct, which teaches a man to avoid the fire; and much as that, which teaches a bird, with such exactness, the art of incubation, and the whole economy and order of its nursery.

Could you answer this classic problem any better?

A problem closely related to Hume's problem is that of underdetermination – this is also a segue into related problems of confirmation. The basic idea of underdetermination of theory by data (or observation) is that the available observational evidence is often not sufficient to decide between rival hypotheses or theories so that evidence for one is simultaneously evidence for the other. Perhaps such rival theories can always be concocted, as the historian and philosopher Pierre Duhem argued? The idea is stronger than it sounds: no possible piece of evidence could ever decide between such theories which are, by construction, "empirically equivalent."

Any observation can be explained in an infinity of ways. Now, science does not, in general, show this proliferation of theories. This shows that other elements are coming into play besides experience: issues of simplicity, economy, and unity might play a role. But these criteria don't come from experience; they must be *a priori* (independent of experience). So it seems that rationalism has entered! Or, if not this, then irrational factors, such as social forces, gender, etc., are in play. To put some flesh on this, consider Copernicus' and Ptolemy's theories just after the former constructed his theory. Recall that Copernicus believed that his model offered a simpler view of capturing the observed phenomena (planetary motions) than did Ptolemy's, which had to postulate motions upon motions (known as "epicycles") in order to account for observations. However, both models were consistent with the data then available so that either model might be said to be confirmed by the planetary motions as then known – it wasn't until Tycho Brahe made new observations that Johannes Kepler was able to go beyond, and indeed step away from, the circular motions of both Ptolemy and Copernicus. We might say both are empirically successful. But they are clearly not *equivalent* in all ways: they make different reality claims.

The problem, to make it plain, is that there are many theories that can account for some empirical observations. There seem to be real historical examples of this, as we will see again in the final chapter. Without some way of selecting a theory, showing how it *uniquely* accounts for the evidence, then what principled reason do we have to select one theory over another? This is basically just another aspect of the problem of induction: such inferences lead us from observed to unobserved instances, but there are multiple ways to get the conclusion (to account for the observed instances). Does the problem of induction, then, force us to claim that all theories are on a par (in terms of justification)? Paul Feyerabend answered this in the affirmative: he argued that the claims of science are not in any way superior to the claims of pseudosciences. Nuclear physics or voodoo, neither is rational: all theories are equally unproven and epistemically on a par. The trust we place in science and the scientific method is therefore totally ungrounded and so totally irrational. This "rationality-gap" leads to the puzzle of why scientists do what they do, why they make the choices they make: if not reason, then what? Enter the historico-sociological explanations of scientific practice, which often point to "mob culture" and "groupthink" to determine theory selection. This leads us back into the problems with pseudosciences: once we reject the idea that science is able to give us access to objective truth, then science loses its privileged status. We return to this, and the problem of underdetermination, again in the very last chapter. Let us now turn to a pair of notorious paradoxes connected to confirmation of hypotheses by empirical evidence.

Confirmation Theory and Evidence

Even the most extensive testing of some hypothesis cannot provide *conclusive proof*. The best one can do is to provide evidential *support* for a hypothesis. This evidential support is known as *confirmation*. In this section we consider two so-called "paradoxes" of confirmation (essentially yet more problems of induction): Carl Hempel's paradox of the ravens (in a series of papers in the journal *Mind* from 1945, entitled: "Studies in the Logic of Confirmation") and Nelson Goodman's "new riddle of induction" (in chapters 3 and 4 of his book, *Fact, Fiction, and Forecast*). Like Hume's problem, these problems have annoyed philosophers of science for many years, and though many solutions have been proposed, like Hume's problem, none seems satisfactory to all.

A major account of confirmation comes from the so-called "Hypothetico-Deductive" (HD) method. The basic idea of the hypothetico-deductive methodology is easy as pie: from some hypothesis H that you are interested in (and given some background knowledge K), deduce a consequence E (for "evidence") that can be checked by observation or experiment. If Mother Nature affirms that E is the case, then H is said to be confirmed or "HD-confirmed." If not (and Mother Nature affirms $\neg E$), then the H is said to be "HD-disconfirmed."

Let's unpack this a bit. A *hypothesis* is supposed to be any statement that is offered up for evaluation in terms of its consequences: we articulate some hypothesis, which can be particular (about *this* thing) or general (about *all* things of this type), from which observational consequences can be drawn. An *observational consequence* is some statement that might be true or might be false, and that can be checked for truth and falsity against observation or experiment. From this we get to the notion of confirmation: if the observational consequence is true, then the hypothesis is *confirmed* to some degree, and if it is found to be false, it is *disconfirmed*.

Let's look at an example, to see this in action. We use the example of "Boyle's Law," from the theory of gases. This states that for any gas (e.g. in a container) that is kept at a constant temperature T, the pressure P is inversely proportional to the volume V:

$P \times V = \textit{constant}$ (at constant T)

Clearly, if we double the pressure on a gas, we will thereby reduce its volume by half. So imagine that the pressure is initially equal to 1 atmosphere (15 pounds per square inch). Now apply a further 1 atmosphere of pressure to the gas, so that the total pressure is now 2 atmospheres. The volume of the gas will then decrease to ½ cubic foot. We can set this up as a hypothetico-deductive confirmation of Boyle's Law:

Boyle's Law: At constant temperature, the pressure of a gas is inversely proportional to its volume
The initial volume of the gas is 1 cubic ft.
The initial pressure is 1 atm.
The pressure is increased to 2 atm.
The temperature remains constant

The volume decreases to ½ cubic ft.

This is a *valid* deduction: the information in the conclusion is already buried within the premises. We have the hypothesis being tested, which is Boyle's Law. In addition, we have further premises that specify "initial conditions" – these are necessary since the hypothesis alone tells us nothing about the world: it doesn't say, for example, whether there is even any such thing in the world as a gas! The conclusion is an "observational prediction" derived from the hypothesis together with the initial conditions. In general, we have the following argument schema:

H = test hypothesis
I = initial conditions

O = observational prediction

Obviously, even if we observe the prediction, we cannot infer the truth of Boyle's Law with certainty: it is perfectly OK to have a valid argument with false premises and a true conclusion (just not true premises with a false conclusion). One can validly infer from the premises to the conclusion, but one cannot reverse this direction to get Boyle's Law as the unique consequence of the conclusion (now taken as the test hypothesis, coupled to the initial conditions). This leads to a serious problem with this account of confirmation: there are *alternative hypotheses* equally compatible with the prediction (that might sit in place of Boyle's Law, for example) – this is also known as the "curve-fitting problem" (since for some finite number of data points, there are many curves one could draw through them, each generated by a different theory). The problem amounts to this: *when an observational result of an HD-test confirms a given hypothesis, it also confirms infinitely many others that are incompatible with the given test hypothesis*. In this case, there are no (empirical) grounds for saying that the test result confirmed one rather than any of the infinitely many other ones, just as with the underdetermination problem mentioned earlier. A further shortcoming is that this method cannot say anything about *statistical* hypotheses: in this case one cannot deduce specific observational consequences, only probability distributions. But the HD method gives no grounds for saying that the premises make the conclusion *more probable*.

Hempel's Paradox of the Ravens

Carl Hempel developed an account of "qualitative confirmation" (i.e. one in which we are not assigning specific numerical values or amounts to confirmations) that provides an alternative to the orthodox HD account. The basic idea is again simple enough: hypotheses are confirmed by their "positive instances." Though this bit is indeed simplicity itself, being pretty much standard inductivism, in order to make it work properly, various *adequacy conditions* have to be met for something to count as a *positive instance* – these are conditions that should be satisfied by any adequate definition of qualitative confirmation. There are four of these (and they, along with this subsection, are a little complicated, so you may have to read through more than once to get it: persevere!):

- *Equivalence Condition*: If evidence E confirms hypothesis H and H is *logically equivalent* to some other hypothesis H', then E also confirms H'.

- *Entailment Condition*: If $E \vdash H$ (E "logically entails" H), then E confirms H.
- *Special Consequence Condition*: If E confirms H and $H \vdash H'$, then E confirms H'.
- *Consistency Condition*: If E confirms H and also confirms H', then H and H' are logically consistent.

On Hempel's account, to take his own example, "All ravens are black" (which we can rewrite as "for any thing, if that thing is a raven then it is black") is confirmed by each individual that is observed to be *both* a raven *and* black. Fairly commonsensical. However, the various adequacy conditions lead to trouble (the paradox) as we see below. The paradox is generated from three simple, quite reasonable-sounding assumptions:

1. If all the *A*s observed thus far are *B*s (i.e. things x that are observed to be A are also observed to be B), then this is evidence that all *A*s are *B*s. [Nicod's Criterion for Confirmation]
2. If e is evidence for hypothesis h, and if h is logically equivalent to h', then e is evidence for h'.
3. A hypothesis of the form "All *non-B*s are *non-A*s" is logically equivalent to "All *A*s are *B*s."

The last two conditions just appear to be basic laws of logic or common sense, and so are pretty much incontrovertible. It is how they interact with the first condition, proposed as a perfectly natural (inductive) way to understand confirmation by the French philosopher Jean Nicod, that causes problems. From these assumptions a paradox quickly follows: all the non-black things we have observed (this white page, your blue shoes, the red carpet, etc.) are non-ravens. So, by invoking the first assumption, we can see that the fact that these non-black things are also non-ravens is positive evidence for the hypothesis that all non-black things are non-ravens. But, using the third assumption, we can see that this hypothesis – "All non-black things are non-ravens" – is logically equivalent to the hypothesis "All ravens are black." So, by the second assumption, we then see that the fact that all of the observed non-black things were also non-ravens is at the same time positive evidence for the hypothesis that all ravens are black. This way one could obtain facts about ravens without ever observing a single bird in one's life! As is sometimes said, one could do indoor (non-bird-based) ornithology in this way. So this is the paradox: observing non-black things, such as my red trousers or my green socks, confirms (i.e. provides evidence for) the theory that ravens are black. Yet surely this isn't the case?

Let's backtrack a little, and flesh out the problem more, focusing on the logic of the situation, which is really the crux. We begin with the simple hypothesis "All ravens are black," which is symbolized in logic as $\forall x\ Rx \supset Bx$ – i.e. for any and all things x (this is the meaning of the upside down A followed by the x), if x is a raven then x is black ($A \supset B$ just symbolizes the conditional statement, "if A then B"). According to a superficially quite reasonable condition (Nicod's condition) a hypothesis is *confirmed* by its positive instances (when we see that the hypothesis is satisfied – so that, in the above case, we find a raven that is indeed black) and *disconfirmed* by its negative instances (equivalent in the above case to finding a raven that was not black, as indeed one might do on Vancouver Island in British Columbia). In other words, the hypothesis that all ravens are black $\forall x\ Rx \supset Bx$ is confirmed by the observation statement "This is a raven and it is black" (in symbols: $\exists x\ Rx \wedge Bx$, where the backwards E stands for "there exists at least one") and disconfirmed by the observation statement "This is a raven and it is not black" $\exists x\ Rx \wedge \neg Bx$.

The paradox comes from sticking to Nicod's condition along with the fact that there are other (equivalent) ways, in logic, of expressing the kind of general hypothesis given above. The point is, logically equivalent statements should be confirmed or disconfirmed by the same pieces of evidence. That seems like a natural condition too. So let's rewrite the hypothesis that all ravens are black in an equivalent way, namely as "All non-black things are non-ravens." This is clearly logically equivalent: if all ravens are black then there can't be a single raven that isn't black, and that means if we find a non-black thing it won't be a raven. In logical symbols we write this equivalent hypothesis as: $\forall x \neg Bx \supset \neg Rx$. This is confirmed by the observation statement "This is a non-black thing and it is a non-raven" ($\exists x \neg Bx \wedge \neg Rx$). By the equivalence condition, this piece of evidence must confirm the original hypothesis that all ravens are black: the same evidence confirms logically equivalent theories. Both hypotheses, being logically equivalent, are confirmed by the same body of evidence, so any non-ravens that are non-black confirm the theory that all ravens are black. Indeed, there are many bits of evidence that would confirm the hypothesis that all ravens are black according to this approach: (1) x is a raven and x is black (the common sense evidence), (2) x is not a raven, (3) x is black, (4) x is not a raven and it is not black, (5) x is not a raven and x is black. This is all to do with the ways in which the statement (the material implication $Rx \supset Bx$) can be true and false: it is only ever false when the antecedent Rx is true and the consequent Bx false, which leaves a lot of other ways of being true! This is clearly a weird consequence: one of the things that is supposed to separate empirical subjects from everything else is the practical engagement with its subject matter.

There are a variety of ways of responding to this problem. Hempel simply bit the bullet and argued that we do indeed confirm the hypothesis that all ravens are black when we observe a non-black thing being a non-raven. It is a psychological illusion that blue books are irrelevant for the hypothesis. Hence, Hempel stoically follows the logic through to its odd end: the hypothesis is really as much about non-ravens and non-black things as it is about ravens and black things. Common sense is just mistaken. So-called Bayesians agree with this idea, bringing in a *quantitative* account of confirmation that is able to make sense of the very tiny amounts of conformation that accrue from such apparently inverted observations: both a black raven and my red sock confirm the law that all ravens are black, but not to the same degree. My red sock offers far weaker support than the black raven. Of course, if this is the case, then scientists have been missing out on a huge chunk of observational evidence. String theorists suddenly have a wealth of empirical evidence. For example, my legs, which are not 6.6×10^{-34}cm long, are also non-strings, thus confirming string theory – where's my Nobel prize for the first empirical confirmation of string theory please?

Goodman's New Riddle of Induction

This particular paradox of confirmation is part of the legacy of the so-called "syntactic view" of theories, which treats them as statements linked by logical relations (to be discussed in chapter 4). Nelson Goodman constructed another confirmation paradox that poses a problem for this idea that confirmation is a matter of logical form. Goodman argues that given any empirical hypothesis, it is possible to devise an alternative hypothesis that is equally well supported by the evidence to date – indeed, there are potentially infinitely many such hypotheses. This means that it is not ever clear which hypothesis is confirmed, and yet both are of "good form" according to the logical accounts of confirmation. Again, this is related to induction, and again is a kind of underdetermination problem. Goodman's problem, also known as "the new riddle of induction" or "Goodman's Paradox," involves both problems of induction (putting a twist on Hume's problem) and problems of confirmation (putting a twist on the paradox of the ravens).

Recall that Hume's problem concerned the question: can inductive arguments give us knowledge (or justification for beliefs gained through induction)? If Hume was right – and despite many attempts to prove him wrong the problem still stands strong – then the answer to this question is No: the kinds of arguments and inferences we use to supposedly give us knowledge of the future, or of things we haven't experienced, do no such thing! At root, the argument is based on a simple matter of logic: while deductive arguments have premises that entail their conclusions, inductive arguments do not. An example of a Humean induction (an inductive argument that faces Hume's problem in an obvious way) is the following:

All *observed* emeralds are green

All emeralds are green

Here, it's obvious that you can't use what's above the entailment line to get what's below it. In cases such as this we are generalizing from some observed instances (say, some object's having a certain property) to all instances *of the same kind*. In other words, if objects x_i had property F every time we observed them (100 xs, say) then we infer that any and all xs have the property F – here, of course, x labels a *type* of object, an emerald, swan, raven, or whatever. Those objects x_j that we are generalizing over (the unobserved instances) might be very far away, or in the distant past, or in the future, or buried deep somewhere.

Goodman's paradox takes both the ravens paradox and Hume's problem further. Whereas the ravens paradox shows that all manner of apparently irrelevant things can confirm general hypotheses, Goodman's shows that we can't even say what a positive confirmation would be in the first place. Goodman's problem says that the format of (Humean) inductive argument is far too liberal: it lets in too many arguments that we don't want to let in. In a nutshell: for many inductive arguments that we would regard as perfectly reasonable (i.e. for which the premises give a good degree of support to the conclusion), there are infinitely many more compatible with the same evidence (the same premises). These infinitely many other arguments are, moreover, completely implausible. To show this, Goodman introduces a new predicate, "grue," defined as follows (with a few modifications from Goodman's original):

x is grue $=_{df}$ either x is green and observed before midnight on July 17, 2019
or x is blue and not observed before midnight on July 17, 2019

This is a bizarre-looking construct, and does not correspond to any properties of objects that we normally encounter – they are not color properties since they involve essential reference to *when* they are observed: two identical colors could be grue and non-grue depending on when they are observed, and conversely, two distinct colors can both be grue depending on when they are observed. We can modify the definition slightly to produce another property known as "bleen":

x is bleen $=_{df}$ either x is blue and observed before midnight on July 17, 2019
or x is green and not observed before midnight on July 17, 2019

To gain some familiarity with this bizarre idea, let's look at an example or two. Let's stick with emeralds since this was Goodman's chosen example, and let me give today's date for me as July 17, 2019 (change the date to your current one, so long as you are before midnight on that date). An emerald (or indeed *any* green gem) that has already been dug out of a rock-face and examined (observed) is grue. Why? Because it must have been examined before midnight on July 17, 2019 *and* it is also seen to be green. However, any green gems that are buried in the rock-face until they are chiseled out and observed on July 17, 2019 are *not* grue. But any blue gems, sapphires and such like, that are chiseled out and examined on July 17, 2019 (and any time thereafter) *are* grue. So, just to get straight on this: an emerald dug up on July 16, 2019, for example, is grue, while one that gets dug up two days later is not. A sapphire that gets dug up on July 16, 2019 is not grue, but one that gets dug up two days later is (see figure 2.6).

Clearly, given this way of understanding things, all the emeralds so far observed are both green *and* grue: they are green, so they satisfy the predicate "... is green" and they have all been examined before midnight on July 17, 2019, so they satisfy the predicate "... is grue" (so satisfying the first bit of the definition of grueness).

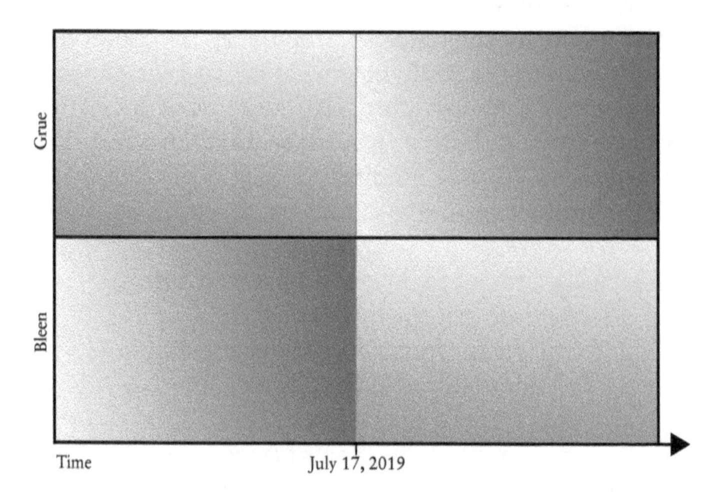

Figure 2.6 A graphical representation of grue and bleen (shade in the top right corner of a block represents blue, with shade in the bottom left representing green)

So all observed emeralds are both green and grue. Now let's get to the problem this poses.

Given that it seems alright to argue inductively from past observations of emeralds being green to all emeralds being green, and given that all observed green things are also grue things, it should be alright to infer that all emeralds are grue too! In other words, we should be able to set up the following inductive argument:

All *observed* emeralds are grue

All emeralds are grue

That is, both the hypothesis that all emeralds are green and the hypothesis that all emeralds are grue are equally well supported by the data. But look what this inductive argument allows us to conclude about future emeralds: it allows us to conclude that any emerald chiseled out of a rock-face and observed after midnight on July 17, 2019 will be blue! Why? Because things which are grue are blue after midnight on July 17, 2019 – that is part of what it is to be grue. But now, given this argument and the one involving greenness, we have a paradox: emeralds dug up in the future (after midnight on July 17, 2019) will be both green and blue (not-green)!

Predicates like "grue" are called *bent* predicates on the grounds that the meanings (the way they are defined) involve a change of direction or a twist (from being green to being blue before and after a certain time, for example). Another name, Goodman's name for them, is *non-projectable* predicates on the grounds that we can't use them for successful predictions (inductive projections of past confirmations into the future). They cause severe problems for induction, since for any nice inductive argument we might come up with (all ravens being black on the basis of many observed ravens being black, and so on) we can come up with an unlimited number of crazy inductions (ravens being "blite," for example, where a blite raven is black before some time *t*, set after all observations of ravens being black, and white thereafter). These can be as crazy as you like; but if the "sane" one is projectable then there is no straightforward reason why the other inductions aren't projectable either. So, if you thought things were bad with Hume's problem, this makes things utterly horrendous! All our nice scientific laws are just as well confirmed as a bunch of bonkers ones.

Just to make plain how serious it is: one can make *any* claim about the future a conclusion of an inductive argument from any premises about the past, just so long as we insert the right gruesome twist in some newly defined predicate. But if this is the case, then the notion of an inductive argument is trivial: it does no work. Moreover, even if we had a solution to Hume's problem, Goodman's would still stand. So, remember (again) that Hume's problem says that the conclusions of inductive arguments cannot ever amount to knowledge (since we can't justify the move from the premises to the conclusion). Goodman's problem says that even if we could solve this problem, so that knowledge from inductive arguments isn't ruled out, we would still not be in a position to say which inductive arguments these are (the ones involving "straight" predicates like "green," or the ones that involve bent predicates like "grue"). For any piece of observational evidence, we can make infinitely many inductive generalizations that are equally in tune with the evidence and yet are incompatible.

So, this isn't just Hume's problem again, nor is it the raven paradox in a disguised form. The concern is not with justifying inductive inferences, but with characterizing those inferences that we do make. What this would involve finding is some asymmetry between "All emeralds are green" and "All emeralds are grue" that makes the former projectable (inductively good) and the latter not.

As soon as we admit that there are additional aspects to confirmation and evidence other than merely observational factors (going beyond empiricism and induction), then the problems evaporate. For example, we might include simplicity, explanatory power, and unification into our notion of evidence. One obvious objection relating to simplicity applies to the predicate "grue" itself (and other bent predicates like it): it looks highly artificial, and cobbled together in a strange way. In particular, it is *disjunctive*, i.e. it contains an "or" in it. We don't expect natural properties to be like this. So the response here is that such properties don't in fact correspond to anything genuinely real in the world. This gives us a way of demarcating between genuine inductive arguments, involving nice un-gerrymandered predicates like "grue" and "blue," and those involving bent predicates – the latter are only "pseudo-inductive" arguments.

But Goodman has a response for this objection, and it involves using the other bent predicate, "bleen," together with grue. With this combination, we can define blue and green, which we don't seem to have a problem with, in terms of grue and bleen:

x is green $=_{df}$ either x is grue and observed before midnight on July 17, 2019
or x is bleen and not observed before midnight on July 17, 2019

This leaves open any property to the same charge of being oddly constructed. A related objection is that grue involves a reference to time, but green doesn't. Again: this falls prey to the inter-definability of green and blue and grue and bleen in the same way. Green, as defined in terms of grue and bleen, *does* involve reference to time.

However, there is another aspect to this which is that the bent predicates, in involving twists after a time, serve themselves up for a "waiting game," whereas straight predicates do not. Thus "All emeralds are green" is better confirmed than "All emeralds are grue" because the latter involves a stage of "further testing." In other words, we are in an "epistemically better" position with respect to straight predicates, so there is an epistemic asymmetry between grue and green, and that is exactly what we need to break the apparent symmetry – according to Bayesians, degrees of projectability are determined by prior probabilities of rival hypotheses against background information, and we assign a greater probability to "All emeralds are green" than to "All emeralds are grue."

Goodman's own response to his problem largely matches Hume's own response to his problem of induction: the reason we tend to speak in terms of non-gruesome properties is simply that they have become "entrenched." Goodman says that "green" is much better entrenched than "grue." Why? Because "green" has been used much more frequently than "grue." Goodman says, whenever we have a situation in which we have some observational evidence equally confirming hypotheses involving bent and straight predicates, the straight one will always override the bent one on account of the entrenchment factor. This is like Hume's answer to the problem of induction because he is giving an account of human usage. He does not go on to say *why* green is more entrenched. So what we have is a kind of conventionalist account of the difference in usages. Given this, his solution is not really a solution at all.

The next section will look at problems faced at the level of the universal generalizations (i.e. universal statements such as "All swans are white"), which fall prey to similar problems as above, when construed overly logically.

Laws of Nature

In rough terms, a *law* is simply a regularity that holds throughout the universe, at all places and all times. We will see that this rough characterization needs supplementing in various ways to avoid problems. We look at three classic views: the *regularity theory*, the *necessitarian theory*, and the *systems view*.

One of the characteristics of laws of nature is their ability to support counterfactuals (i.e. statements of the form "If *A* had happened, then *B* would happen") and modal statements of necessity and impossibility (i.e. about what *must* happen and what *cannot* happen). Suppose I have some salt and a glass of water. Then it is true that *if I were to put the salt into the water it would dissolve*. This is true even though I do not in fact do so. The statement in italics is an example of a counterfactual. The counterfactual can be truthfully asserted because it is supported by a law involving the solubility of substances in water. Laws are thus supposed to transcend their actual instances. The same cannot be said for the beer in my fridge example: if I have a bottle of Theakston's *Old Peculiar* beer in my hand, I cannot claim (on the basis of the true universal regularity that all the beers in my fridge are Timothy Taylor's *Landlord*) that it would become a bottle of Timothy Taylor's *Landlord* if it were in the fridge. This is what is meant by laws supporting counterfactuals: they are supposed to have a special status in the universe, and are often thought to be essential for our being able to do science at all.

Laws of nature also support modal claims involving the impossibility or necessity of certain things (impossibility is the same as being *necessarily* not possible). For example, it is a regularity of nature that nothing transmits information faster than the speed of light. But it is more than this: there is a law *forbidding* such transmissions. Hence, it is impossible that information can be transmitted faster than the speed of light because of the laws of physics (the laws of special relativity). The laws in this way support modal claims. It is a regularity that no humans travel at half the speed of light: this is a true fact about the universe. But this is not sufficient to ground modal claims about impossibility: it is *possible* that a human be propelled at half the speed of light, provided sufficient energy be provided for the task. This keys into another distinction: that between contingency and necessity. It is only a *contingent* fact that no human travels at half-light-speed: meaning precisely that it is physically possible (in a way consistent with the laws of physics) for humans to travel at such speeds (though they may not be recognizably human at this speed!). However, it is *necessary* that nothing travel faster (or transmit information) faster than light speed. Alternatively, there is a distinction between "accidental" and "lawful" generalizations – this better describes the situation with the beer in my fridge: it is a mere accident that they are all one brand, rather than being that way because of the laws of the universe. But how are we to make sense of these curious entities, laws of nature?

David Hume (in *Treatise on Human Nature* (§1.3.14 and §1.3.15)) defended a "regularity theory" (aka the "Humean Theory"): laws of nature are nothing but *true universal generalizations*. Let's give the standard example: "All metals expand when heated." And the regularity theory explanation of this? All pieces of metal that are heated expand. According to Hume, this is simply the correct empiricist view of laws, and it played a direct role in the Logical Empiricist's account of scientific theories and Hempel's account of scientific explanation as we see in the next section. In summary: *there's nothing more to laws than what actually happens in the world.*

There are many problems with this account, some of which we have alluded to already. Chief amongst these is the problem of *vacuous laws* (a problem raised by philosophers Fred Dretske and Hugh Mellor). Recall that we write laws as $\forall x\ Fx \supset Gx$ (for any thing x, if x is an F then it is also a G) – e.g. "all metals expand when heated" $\equiv \forall x\ Mx \supset Ex$ (here, "M" stands for "is a metal" and "*E*" stands for "expands when heated"). But, as we have seen when looking at the ravens paradox, $\forall x\ Mx \supset Ex$ is logically equivalent to $\neg\exists x\ Mx \wedge \neg Ex$ (i.e. it is not the case that there is a thing that both is a metal and that doesn't expand when heated) – this means that if there were no metal at all in the universe then the law would be *trivially* satisfied. The regularity theory just says that laws are true generalizations, so whenever the antecedent of a law statement has no instances in the world (such as when there just is no metal), the law is true: but this is a vacuous type of law. It is too easy to construct them: "all unicorns travel at light speed," etc. A common sidestep out of this problem is to try adding additional elements, making sure that what the law is talking about actually exists (an *existential condition*): "All *F*s are *G*s" iff $\forall x\ (Fx \supset Gx) \wedge (\exists x\ Fx)$. This says that there *are* objects of the type specified in the antecedent slot of the law statement. Since there are no unicorns, the law that all unicorns travel at light speed is no longer rendered true. But this leads right into the jaws of another problem involving "non-instantial laws." The idea here is that the existential condition is just too strong since it rules out many standard examples of what we surely ought to consider *good* scientific laws. For example, Newton's first law of motion states: "All bodies on which no net external force is acting either remain at rest or move at uniform velocity in a straight line." If the existential condition is true, then this does not count as a law! One of the most famous laws cannot be a law. The reason? All bodies exert a gravitational force on each other (however small), so no bodies are ever free from external forces. Therefore, Newton's law fails to satisfy the existential condition, which is absurd.

The intuition is to think of uninstantiated laws in terms of how objects *would* behave if they did exist. So: if there were bodies *on which no net external force is acting*, then they would *either remain at rest or move at uniform velocity in a straight line*. But this move is not open to the regularity theorist: they ground laws in what *actually* happens in the world, not what *might* happen in other "possible worlds" in which things are different. Even if this problem can be solved, for laws like Newton's, the problem still affects so-called "functional laws," which describe a functional relationship between two or more variables (i.e. in the form of a mathematical equation) – that is, they tell us how one or more (dependent) variables change their values as a result of changes in other (independent) variables' values. An example is the ideal gas law that we have met before: *P*(*ressure*) × *V*(*olume*) = *nR* × *T*(*emperature*) ("*R*" is the "gas constant") – what this says is: "the pressure times the volume of *n* moles of gas is proportional to the absolute temperature of the gas." The problem with such laws is that the magnitude of the variables involved ranges over an infinite number of values (e.g. the interval between 0°C and 1°C contains infinitely many possible values), but only a small finite number will ever be realized: no gas will be heated or cooled to all possible values of *T*. We face a similar problem to the preceding one: the law still tells us what the pressure of a gas would be if its temperature were 1 trillion degrees. This is never going to happen: so we seem to be forced into *counterfactual* territory again (and *out* of empiricist territory) – we need to say what *would* happen to a gas's pressure if we heated it to 1 trillion degrees. We seem to be pulled away from an account of laws involving only what actually happens; though the determined Humean can simply dig their heels in and deny any special status to laws, the onus is on them to then explain the success of science in light of their deflationary view.

In fact, the previous two problems get worse since, according to regularity theorists, no law can refer to any un-actualized possibilities whatsoever. So, the problem is this: if "something is *F*" (or "*Fx*") is a statement of an unrealized possibility (i.e. something that might happen but in fact hasn't), then it is false. But if it is false, then the regularity view turns it into a law of nature: "Nothing is *F*" (or "$\neg\exists x\, Fx$"). But now this means that "something is *F*" is inconsistent with a law of nature, so it cannot be a statement of an unrealized possibility after all. To see how silly this is, think of what it implies: either something is true or else it is impossible! If the silliness of this is not evident: had the Americans not dropped an atom bomb on Hiroshima then there would be a law of nature preventing it from happening. Or, as philosopher Bas van Fraassen nicely expresses it, it is also a law of nature that there cannot be a river of Coca Cola since there are in actuality no such rivers. Perhaps the most serious problem is one we have touched upon already, that of accidental regularities that are not laws: there are many true universal regularities in the universe (e.g. "all the beer in my fridge is Timothy Taylor's *Landlord*" – there has *never* been a beer of a different type in my fridge). But not all such regularities can be laws: it is not *impossible* to put another kind of beer in my fridge; there is no force that pushes them out! And being a bottle of beer in my fridge does not *imply* that it is a Timothy Taylor's *Landlord*. So only *some* regularities appear to be lawful, and others do not: these other regularities are merely "accidental." But how to distinguish between these using only the regularity account? There's simply no way to do it without appealing to counterfactual cases and/or possible worlds. The classic example here is the following (again due to Bas van Fraassen). Consider the following statements:

1. All solid spheres of gold weigh less than 100,000 kg.
2. All solid spheres of pure plutonium weigh less than 100,000 kg.

The first is not a true statement of a law: there might be some worlds with laws of physics and chemistry exactly like ours, in which there are gold spheres with masses greater than 100,000 kg – perhaps some civilization travels through the universe collecting gold to make such a sphere. However, the second statement is true: plutonium would be unstable for masses much less than 100,000 kg (assuming the density is not too low). Note, however, that both are true of our universe: there are spheres of neither type for these masses. But the latter is true in a way the former is not. Universality of form – i.e. being able to put the generalization into the form All a's are b's – is not sufficient for being a law. What we need to get this distinction is the notion of a "counterfactual conditional": "if a had been the case, then b would have been the case." If we run the two statements above through this, then we see that only the plutonium case is true:

1. If there were a solid gold sphere, then it would weigh less than 100,000 kg.
2. If there were a solid pure plutonium sphere, then it would weigh less than 100,000 kg.

The reason the latter is true and the former false, despite the identical form of the statements, is that the latter has a supporting law concerning the critical behavior of plutonium masses (see figure 2.7). The regularity theory is unable to distinguish these cases. Several philosophers tried to point to some element that must be added to the regularity theory to allow for a distinction to be made, if only in the subjectively distinct attitudes that we have towards them that leads us to more readily use cases like the plutonium example in predictions or scientific explanations. But note that such modifications would leave many of the other problems with the regularity account undisturbed.

Most other views of laws of nature take their cue from the problems with the regularity theory. The most famous alternative is probably the *necessitarian view*. This admits that the regularity theory, with its empiricist backbone, cannot capture about laws what we want

Figure 2.7 A 6.2 kg sphere of plutonium (surrounded by neutron-reflecting blocks of tungsten carbide) known as "the demon core." Such spheres cannot, as a matter of the laws of nuclear physics, be too massive, unlike spheres of gold

Source: reproduced with permission of Los Alamos National Laboratory [Contract No. DE-AC52-06NA25396 (US Department of Energy)]

it to capture, namely their modal and counterfactual power. Necessitarians admit that laws of nature are not contingent matters of fact that hold in our world, but are physically necessary: at least some notion of necessity, some modal machinery, is required to make sense of laws of nature. The first account of this kind was supplied by David Armstrong, Fred Dretske, and Michael Tooley. The basic idea is that laws of nature concern properties and the connections between them. The properties are taken to be *universals* ("red" is a universal since there can be lots of different particular red things at the same time in different places: red fire engine; red rose; red underpants, etc.). Laws are then of the form "*F*-ness yields *G*-ness," where something's being *F* *necessitates* its being *G* in virtue of the relation between the universals *F* and *G*. Newton's law $F = ma$ is a law, and on this account we read it as: "the properties of being subject to a force *f* and having a mass *m* necessitate the property of accelerating in the direction of this force at $\frac{f}{m}$ *meters/second*^{2}." So we have the view that laws are necessitation relations *N* between pairs (or greater multiples) or universals *F* and *G*, or *N*(*F*,*G*). So, this relation *N*(*F*,*G*) implies and explains the universal regularity connecting *F*s and *G*s, but not *vice versa*. The necessity is in the world of the universals (of properties of things) rather than of objects themselves: the objects may "instantiate" these universals and then the "mustness" of the law is passed on to these particular objects, as is the case with the plutonium sphere which is forced to obey the criticality laws that are imposed by the properties of nuclei. This setup allows necessitarian theorists to avoid the problem of accidental regularities almost by *fiat*: laws support counterfactuals in virtue of a pre-existing relation between universals – something lacking in the accidental cases. Not surprisingly, there are problems. Not least is the task of explaining what is this relation that can do this remarkable thing of gluing together disparate properties as we find them expressed in the laws of nature. All its architects seem to have done is give this relation the name "necessitation" in the hope that it will somehow be enough.

The so-called "best systems" account of laws is an empiricist reaction to what it perceives as the metaphysical excesses (universals and strange universal glue!) of the necessitation theory: it sticks firmly to the occurrent facts. The basic idea, due to David Lewis but with various precedents, is to view laws of nature as axioms or theorems that live in those deductive systems that best balance *strength* and *simplicity*: strength here means the amount of *information* the system encodes about the world and simplicity refers to the efficiency with which the system organizes the mass of diverse facts about the world. The idea is that there is nothing metaphysically fishy about all this: empiricists can use this definition and not worry about the problems faced by the regularity theory. The law concerning the plutonium sphere would be included as a law on this account since it derives (i.e. can be deduced as a theorem) from quantum theory, which would obviously be amongst our best systems for the world. Vacuous laws, like those describing the properties of unicorns, would not be derivable from the best systems, and so would not be genuine laws. Not allowing uninstantiated laws (the other major problem of the regularity theory) would make the system's strength suffer. The main problems with this account might by now be apparent to the reader: "strength" and "simplicity" are highly subjective. The notion of "information" too is not without its problems. Have we not now just traded the problem of laws for this other problem of making sense of these new concepts?

In the next section, we turn to a topic that includes laws in a central way, namely scientific explanation, and in fact the regularity theory was explicitly involved in a prominent approach, and faced many similar problems to those we have seen already with confirmation and evidence too – again the source is the basis in a specific way of formulating concepts using logic.

Models of Scientific Explanation

The question "Why?" is often asked in order to seek *understanding* of something: some event, action, or phenomenon. In this section, we are interested in why-questions asked in the context of science, or with regard to scientific phenomena. The understanding sought will then be scientific understanding. This understanding will be delivered in the form of an *explanation*. Our job in this section is to understand what a scientific explanation consists in. Confirmation looks superficially similar to explanation in many cases. However, whereas confirmation concerns reasons for believing *that* a given phenomenon occurs, explanation is concerned with *why* the phenomenon occurred. So, take the hypothesis that all emeralds are green. Confirmation will involve the observation of many emeralds and noticing that they have all been green, which lends support to the general hypothesis (sweeping Goodman's grue paradox under the rug for the moment!). The explanation of this hypothesis, on the other hand, will involve saying why emeralds are green. This might involve an analysis of the chemical composition, the specific crystalline structure, and the interaction of light with this structure. However, just as confirmation was cashed out in terms of an argument, so too is explanation on some accounts: we look at the most famous, Hempel's "covering law" or "deductive-nomological" model first. As the name suggests, laws of nature play a central role.

Hempel's DN ("Covering Law") Model

In his paper "Studies in the Logic of Explanation," written with Paul Oppenheim in 1948 (*Philosophy of Science* **15**(2): 135–75), Hempel devised an account of scientific explanation known as the "Deductive-Nomological Model" (more popularly known as "the Covering Law Model," for reasons that will become clear below – sometimes known as the "Subsumption Model" for the same reasons). This model conceives of scientific explanations as arguments given in response to "explanation-seeking why questions." According to this model, a scientific explanation is an argument consisting of two parts:

- An *explanandum*: a statement "describing the phenomenon to be explained" (the "conclusion").

- An *explanans*: "the class of statements which are adduced to account for the phenomenon" (the "premises").

In other words, the *explanans* work together to give reasons for the *explanandum*. Or, in still other words, the *explanans* is the bit that does the explaining, while the *explanandum* is the bit that needs explaining: the former leads us to *expect* the latter. Or, in more words, a scientific explanation is an argument whose conclusion is the statement expressing the phenomenon that requires explanation and whose premises say why the conclusion is true.

To give a scientific explanation for why objects look bent when immersed in water, we have to construct an argument whose conclusion is "objects look bent when immersed in water" and whose premises tell us why this is so. What is missing is an account of the relationship between premises and conclusion that has to hold for the argument to count as a genuine scientific explanation.

To accomplish this task, Hempel's Deductive-Nomological Model of explanation required the following four individually necessary and jointly sufficient conditions:

(i) The explanation must be a valid deductive argument (i.e. the premises should *entail* the conclusion).
(ii) The sentences (propositions) in the *explanans* must be true.
(iii) The *explanans* must be empirically testable.
(iv) The *explanans* must contain at least one general law (such as "all metals expand when heated") that is actually needed in the deduction of the *explanandum* fact.

So, the explanation should be a *sound* deductive argument: the premises (i.e. the *explanandum*) should be true, and the premises should entail the conclusion (the *explanans*) – this much gives the "deductive" part of the definition of explanation. Any explanation satisfying these conditions gives enough information to predict the *explanandum* fact given the initial (or boundary) conditions. Note also the appearance of general laws (or "laws of nature") in the list of conditions: this is where the "nomological" part comes from (it is also why this model is also called the "covering law model": the law *covers* the phenomenon to be explained, which might be a particular fact or a general law). Though particular facts can play a role in explanations too, Hempel believed that a general law was essential.

To sum up, then: to explain something on Hempel's account involves showing that something's occurrence follows deductively from a general law (or from general laws), along with other particular facts (i.e. facts that refer to particular things, times, places, etc.), which must all be true. Schematically, we have the following:

General Laws L_i
Particular Facts P_j

Phenomenon to be explained

As mentioned above, the "thing to be explained" (the *explanandum*) can be either a particular fact (why *my* leg looks bent immersed in water) or a general law (why objects look bent when immersed in water). The general laws here would be, for example, the laws of optics. Particular facts would be such things as the angle of elevation of the Sun and so on. It should be obvious why this is called the "covering law model" now: the general law(s) have the thing to be explained "covered." This is just what an explanation amounts to: find a law that covers the phenomenon.

Hempel was not ignorant of the fact that there were cases of scientific explanations that, on the surface, look free of general laws. Michael Scriven gave the very simple example of an inkwell being knocked over by someone's knee. Here, the thing to be explained is the falling of the inkwell. The explanation is simply the knee's knocking it off! No general laws here it seems. But Hempel maintained that if one were to spell out in full detail this situation, then it would involve laws: the knee's knocking into the inkwell would be covered by some general laws involving, e.g. biomechanics (for the knee jerk!), mechanics, gravity, condensed matter physics, and so on. It wouldn't be a pretty explanation, but this would be the correct one.

Hempel was able to draw an interesting consequence from his analysis of explanation: prediction is just an aspect of explanation – they are two sides of the same coin. If the *explanandum* hadn't already been observed, we would have been able to predict it with exactly the same argument. If we hadn't already known about the bent appearance of objects immersed in water, we could have predicted the bent appearance of my leg in water before immersing it, by simply employing the appropriate law and inputting the appropriate values. Explanation and prediction are, in this, symmetric. Making a prediction – say that the hole in the ozone layer will have doubled in ten years – will serve to explain that fact after it has happened.

Elegant though the account is, and though the account gets much right about scientific explanation, it faces many counterexamples that prove fatal. There are two classes of counterexamples: (1) those showing that Hempel's account is not *necessary* for scientific explanation (by finding an explanation that counts as a "genuine" one but that doesn't fit the covering law model), and (2) those showing that Hempel's account is not *sufficient* (by finding things that blatantly are not genuine scientific explanations but that fit the covering law model all the same). So, Hempel's model comes out as both too strong (for ruling too much out) and too weak (for ruling too little out)! Let's briefly present these in order.

First, then, we are concerned with finding examples of arguments that do not fit Hempel's covering law model, and so do not class as good scientific explanations on his account, but which are clearly perfectly good explanations. This shows that the DN account is not necessary: it is too strong or "restrictive." I already mentioned Scriven's example of the inkwell. The idea of this case is to highlight an instance in which we have a (singular causal) explanation (the impact of my knee on the desk caused the tipping over of the inkwell) of some event without any general laws appearing, thus violating one of Hempel's necessary conditions. The problem is, this looks like a clear case of explanation. Hempel's way out of this is to say that there is an underlying explanatory structure to the above singular causal claim that fits the covering law model. There will be a general law about knees (in general) striking desks with inkwells on them (in general) in certain ways. There will be initial conditions stating that a knee struck the desk in just the way necessary to tip the inkwell over. There will be facts about gravity and so on. It's perfectly true that we can do this I suppose, but the question is: does the singular causal claim *require* that there be this underlying structure in order to be explanatory? There doesn't seem to be a reason why there should be.

What about examples of arguments that fit Hempel's covering law model, and so qualify as good scientific explanations on his account, but which are clearly not explanations at all? This shows that the DN account is not sufficient: it is too weak or "liberal." The classic example of this kind is due to Sylvain Bromberger. Suppose you notice that a flagpole is casting a shadow of 20 meters on the ground (see figure 2.8). You are asked to explain why the shadow is this long. This is the kind of question that Hempel would accept as beckoning a scientific explanation that should fit within his covering

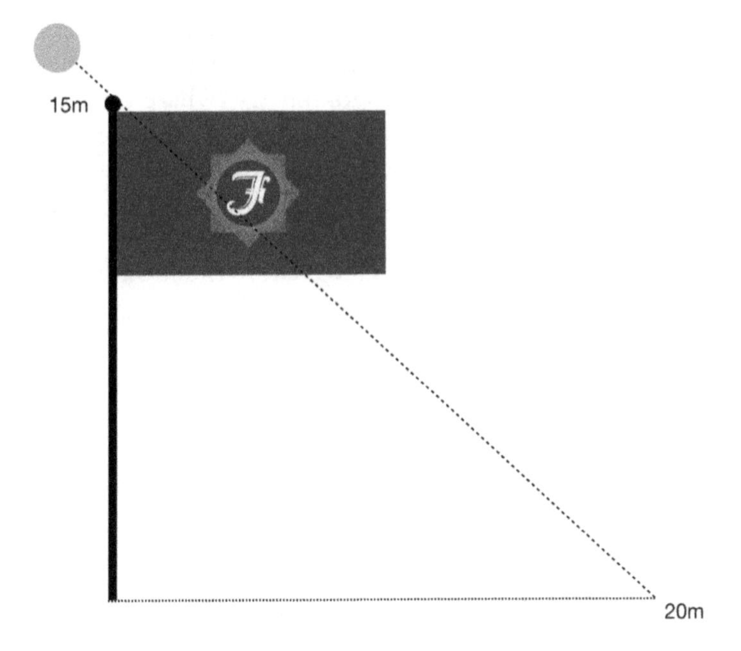

Figure 2.8 The flag of Freedonia. Here, the shadow a flagpole casts can be deduced from laws of optics and the height of the flagpole, thus providing an explanation of the shadow's length. However, the situation is symmetrical, allowing the flagpole height to be similarly deduced from the length of the shadow and the laws of optics

law model (it is an excellent example of an explanation-seeking why question). If you were a clever sort you would be able to answer by mentioning trigonometric and optical laws, and some facts about the flagpole. For example:

(i) Rays of light from the Sun are hitting the flagpole
(ii) The flagpole is 15 meters high
(iii) The angle of elevation of the Sun is 37°
(iv) Unimpeded light travels in straight lines
(v) By trigonometry *tan* 37° = 15/20 gives us 20 meters for the length of the shadow

A quick inspection shows that this fits Hempel's model perfectly: the general laws are the laws of optics (straight-line travel of light) and of trigonometry (where we used a simple formula for getting the length of one side of a triangle – that corresponding to the explanandum-fact – from an angle and another length); the particular facts (or initial conditions) are the flagpole height and the angle of elevation of the Sun. We can write it out as an explicit general argument:

(L1) Light travels in straight lines
(L2) Laws of trigonometry
(P1) Angle of elevation of Sun
(P2) Flagpole is 15 meters high

(C) Shadow is 20 meters long

The premises are true, the argument is deductively valid (once we specify the laws in finer detail), and the conclusion follows (it is true too). So far so good: what's the problem? The problem is we can simply switch the *explanandum* (C) (the shadow being 20 meters in length) with P2 (the flagpole's being 15 meters high), and get an argument that also fits Hempel's model for a good scientific explanation. The trouble is, it now looks as though the shadow of the flagpole is explaining the height of the flagpole:

(L1) Light travels in straight lines
(L2) Laws of trigonometry
(P1) Angle of elevation of Sun
(P2) Shadow is 20 meters long

(C) Flagpole is 15 meters high

You might make a case for this still qualifying as "sensible": perhaps the shadow length *was* responsible for the flagpole being a certain height; maybe the shadow has to cover a certain length for ceremonial reasons (the Aztecs played around a lot with this kind of thing). But it gets worse: we can also swap around P1 with C!

(L1) Light travels in straight lines
(L2) Laws of trigonometry
(P1) Flagpole is 15 meters high
(P2) Shadow is 20 meters long

(C) Angle of elevation of Sun

This is again a perfectly valid and sound (the bits are all true) argument: it satisfies Hempel's model. But now it looks like the shadow and the height of the flagpole are explanation of why the Sun is where it is in the sky!

This is why Hempel's account is too liberal, too weak. The lesson exposed by this problem is that explanation is an *asymmetric* notion: if x explains y, then it is not, in general, going to be the case that y explains x. Hempel's account does not capture this directedness. This problem also causes problems for Hempel's claim that explanation and prediction are two sides of the same coin: if you hadn't already seen the angle of elevation of the Sun, then you would be able to predict it from the laws and from the height and length of the flagpole and the shadow it casts. But you would not then say, after you observe the predicted result, that this *explains* the result of the sun's being where it is. Prediction and explanation are not the same after all. As another example of this, note that given the present positions of the planets together with the laws of classical mechanics, astronomers can predict the future positions, including solar eclipses. Once the eclipse has occurred, the data, laws, and computations provide an explanation of the eclipse. But one can also retrodict previous eclipses from the same data and laws. This fits DN too. But we don't want to say that this present state *explains* the earlier eclipse: that seems like meddling with the direction of time.

Another related problem is that of the common cause. In this case, let the law be that storms follow a drop in the barometer reading. Suppose that the barometer drops, and there is indeed a subsequent storm. This follows the DN pattern too. But we don't say that the barometer dropping explains the storm. Why not? Because it doesn't play a *causal* role in it: they are both effects of a common cause, the drop in atmospheric pressure. Hempel's model has serious trouble with causal relevance. The best example of this is the case of the male birth control pill. Suppose a child naively asks why their father, John, doesn't get pregnant. Consider the following argument:

(L) People who regularly take birth control pills do not get pregnant
(P) John has been taking his wife's birth control pill regularly for a few years
----
(C) John does not get pregnant

Again, this fits the covering law model perfectly well (we'll assume P is true), but it is clearly quite mad: the fact that John has been taking the pill is completely irrelevant to his not getting pregnant! Clearly this argument is no explanation. The correct argument, with the correct law, should be:

(L) No male gets pregnant
(P) John is male

(C) John does not get pregnant

This yields a good explanation, but Hempel's account cannot distinguish between such bad and good arguments. It is not sensitive to what is relevant and what is not. Explanations should have this sensitivity, so Hempel's model misses out an essential component of explanation. Again, the covering law model is found to be too weak in letting in as explanations things we wouldn't want to class as such.

The covering law model explains things by deducing them from *deterministic* laws (i.e. those with unique outcomes) and initial conditions. Many laws in science are statistical – those in quantum theory, Mendelian genetics, and so on. So the question is: do these statistical laws figure in explanations too? Surely they do, yet it cannot be along the lines of the DN model for, as we know from our tales of induction, we can't get a unique conclusion – the DN model cannot be the whole story, even if we ignore the counterexamples. For example, there might be a probability for contracting some disease, but this isn't enough to determine that someone *will* get the disease. So, given an explanation-seeking why question such as "Why did John get lung cancer?," we might respond by invoking some statistical law connecting smoking to lung cancer (such as 30% of male, habitually heavy smokers between the ages of 40 and 50 contract lung cancer), and then filling in the initial conditions, that John *was* a heavy smoker, and was 45, etc. But we cannot predict that John will get cancer since this is probabilistic. We can at best predict a probability that John will get cancer. This pattern of explanation Hempel called "Inductive-Statistical" (IS): individual events are, in this case, covered by statistical laws. One cannot deduce that John will get lung cancer even though all of the premises are true (John does smoke heavily, etc.). In IS explanations, the inference from premises (*explanans*) to conclusion (*explanandum*) is inductive rather than deductive (hence "Inductive-Statistical"). An IS explanation is good if its *explanans* confers high probability on the *explanandum*.

The IS model attempts to explain particular occurrences by subsuming them under statistical laws. Let's use Hempel's own example. Suppose we are interested in the rapid recovery of some patient. We ask the (explanation-seeking why) question "Why did John Jones recover quickly from his streptococcus infection?" The answer is that he was administered a dose of penicillin, and strep infections usually clear up in such cases. We can write this out as:

(L) Almost all cases of streptococcus infection clear up quickly after the administration of penicillin
(P1) John Jones had a streptococcus infection
(P2) John Jones received treatment with penicillin
---[prob]
(C) John Jones recovered quickly

Here, the inference is inductive rather than deductive, and the "[prob]" at the end of the inference line indicates that the *explanans* support the *explanandum* with a certain probability (whatever probability the statistical law – here, involving recovery with penicillin – confers). Notice that though this is inductive, we still have the connection between explanation and prediction. Had we not known of John's recovery, we could have predicted its outcome with a certain probability.

Hempel noticed a problem with the IS account that he labeled the "problem of ambiguity of IS explanation." The problem can be spelled out as follows: suppose that in addition to having a strep infection, it was also noticed that John Jones had a penicillin-resistant strain of strep infection. Penicillin-resistant strains do not lead to quick recovery on the administration of penicillin. We have now the following argument:

(L) Almost all cases of streptococcus infection clear up quickly after the administration of penicillin
(P1) John Jones had a penicillin-resistant streptococcus infection
(P2) John Jones received treatment with penicillin
---[prob]
(C) John Jones did not recover quickly

Now, the premises of both arguments are consistent: they could all be true. But their conclusions are not; they directly contradict. Hempel sought to overcome this problem – that what appear to be good IS explanations can be invalidated in an instant – by introducing "the requirement of maximal specificity": include all relevant knowledge when constructing IS explanations. Had we known about John's resistant strain, the first argument would not have been an acceptable IS explanation. By adding this requirement to his list of adequacy conditions given earlier, we have conditions that must be satisfied by *all* explanations. This theory of explanation was for a long time "the received view." There are counterexamples to the IS account too, some following the same kinds of line as for the DN model. Here we just take a problem that differs from those afflicting the DN account, known as "the paresis problem."

Paresis is a form of tertiary syphilis: it is contracted only by people who have gone through other stages of syphilis without receiving treatment with penicillin – it is truly the stuff of nightmares. So, one might ask "Why does John suffer from paresis?" (poor old John!). The answer (the bit in the *explanans*) would be that John had a case of syphilis that went untreated. The problem with this is that only a small percentage of those with untreated syphilis actually go on to contract paresis: 25%. So, if you have a roomful of people with untreated syphilis, only one in four would go on to develop paresis. Hence, on the basis that someone has untreated syphilis, the correct conclusion to draw is that they will *not* go on to develop paresis. But the existence of untreated syphilis was used in the *explanans* for John's contracting paresis. The problem with this is that it does seem like a legitimate explanation for why John has got paresis: people don't spontaneously get paresis; they *have* to go through various stages of untreated syphilis. So it seems like a good explanation, but it isn't according to Hempel's account since the premises render the negation of the conclusion more probable (i.e. it's more probable *not* to contract paresis). The point of this problem is that untreated latent syphilis is certainly relevant to the contracting of paresis, but it does not make it highly probable (in fact, it makes it less probable than *not* contracting it).

It is hard not to have the feeling that statistical explanation just masks the fact that we don't have complete knowledge of the workings of the world. The reason we give a statistical explanation of the interactions of strep infections with penicillin is that we don't know which specific people will recover. Likewise, the reason why we can only say a quarter of the people with untreated syphilis will get paresis is because there are facts we don't know causing those that do get it to get it. There is surely always a story like this. If we could look at the detailed micro-laws of such situations, we would surely be able to give a DN explanation. This would mean that the IS style of explanation is done purely for convenience as a result of our limitations: statistics emerge through ignorance. But then we face all of the original problems facing the DN model.

The covering law model is unanimously agreed to be fatally flawed. Most accounts of explanation that followed Hempel's account are largely reactions to it, attempts to avoid the problems that plagued it, often by basing the account precisely in this avoidance. One suggestion for an alternative is to use causation to ground the notion of explanation: one explains a phenomenon by citing its cause(s). This seems promising: the flagpole and pill problems appeared to be problematic precisely because there was no causal link flowing from shadow to pole or Sun and from taking the pill to not becoming pregnant. It also seems to match an awful lot of what goes on in science: if we find a species that went extinct, we look for its cause (meteorite, change in the environment, extinction of prey, etc.). Causation is also asymmetric, so we don't get the symmetry problems. But it conserves many features of the covering law model also. Like the covering law model, a phenomenon that we wish to understand is being deduced, to a certain extent, only this time explicitly from a cause: this cause might well still involve laws in addition to initial conditions.

One problem with this account is posed by "theoretical identifications": situations where two things (or concepts) from distinct theories are identified. Examples are: "Water is H_2O" and "heat is mean molecular kinetic energy." This is a problem for the causal accounts because we wish to say that we have explained heat when we say that it is just average molecular kinetic energy, but we do not wish to say that average molecular kinetic energy *causes* heat: it just *is* heat. If your intuitions aren't clear with this example, use the water and H_2O example instead: H_2O does not cause water, it *is* water!

Wesley Salmon, an American philosopher of science, likes Hempel's account, but is keen to avoid the pitfalls to do with symmetry, causation, and irrelevance. In other words, he wants his account to be asymmetric and sensitive to the difference between a causally (explanatorily) relevant factor and a non-causally (non-explanatorily) relevant factor. He does this by introducing the notion of "statistical relevance" which is a relationship of conditional dependence between properties, somewhat like causation, but broader – this is explained in his book *Statistical Explanation and Statistical Relevance* (University of Pittsburgh Press, 1971). This account does not involve the idea that explanations are arguments. Salmon is concerned most with statistical explanation. The counterexamples to the IS model involved the notion that high probability isn't really important; what is important is statistical relevance. Statistical relevance is a comparative idea involving the relation between different probabilities, that of the hypothesis alone $P(H)$ and that of the hypothesis *given* the evidence $P(H|E)$. If the evidence is statistically relevant, then the probability of the hypothesis is raised or lowered by the evidence; otherwise it is the same. The basic idea is very simple, though the notation may be unfamiliar: if a factor makes a difference to the probability of another factor, then that factor is statistically relevant (positively or negatively). It doesn't matter whether some statistical law confers high probabilities (as in the paresis case): all that matters is that the law makes some difference to the probabilities for some outcome or event.

The main problem with this account is that, although it involves the notion of statistical relevance, and so avoids certain problems in Hempel's model, it does not really say anything directly about causation. This is a problem because the statistical relevance account leads to a mixing up of causation and correlation. Two things may be correlated (so one's value changes when the other's does), and yet it might not be the case that either is a cause of the other: they might both be effects of some common cause. Think of the barometer example again: here, the barometer and the storm are correlated in a way that would make one statistically relevant for the other. We might write: $P(storm|barometer) > P(storm)$ (that is, the probability of a storm occurring given the barometer drop is greater than without a barometer drop: it is statistically relevant in Salmon's proposed sense. The barometer's dropping indeed increases the probability of a storm coming. But the drop in the barometer reading does not *cause* the storm. Therefore, it doesn't make sense to say that the dropping barometer reading caused the storm: both are the result of the drop in atmospheric pressure which is the common cause.

It is clear, also, that the statistical relevance account is really the search for causal factors of phenomena. When we do a randomized trial to check for the statistical relevance of vitamin C on recovery from the common cold, we are checking for causal relevance: does vitamin C have an *effect* on recovery? Does it make a difference? Clearly causal relevance, rather than statistical relevance, has the explanatory weight: statistical relevance simply piggybacks on causal relevance. We find it useful to use statistical relevance as a way of making inferences to causal relevance, but it is causal relevance that does the work.

Paul Humphreys has a proposal similar to Salmon's in that explanations involve the citing of causes that affect the probability of some effect (the explanandum-fact; the thing that needs explaining) – see his book *The Chances of Explanation* (Princeton University Press, 1989). Humphreys adds to this that the impact on the probability of the effect brought about by the cause is "invariant": regardless of how you mess with things other than the cause (i.e. with the background conditions), the phenomenon of interest still occurs. So: explanations cite causes, and causes have to invariantly modify the probability of occurrence of some effect. That is, regardless of how external conditions are altered, the probabilistic relation stays the same, and this is necessary for causation to be seen as operating, and for the explanation involving the cause to be kosher. Why is this? Because explanations involving probabilities (statistical explanations) generally concern situations where the background conditions are very different: John who smokes 40 cigarettes a day also lives in a well-to-do area, with little smog, etc.; contrast this with John who smokes 20 a day, but lives in a poorer area, is smaller, weighs less, etc. In both cases we want to say that there is a law connecting cigarette smoking and lung disease, and this holds regardless of the variations in background conditions. The problem with this way of understanding explanation is that no cause will be able to figure in a scientific explanation unless we can show that its modification on the probability of the effect remains the same under *all* possible alterations of the background conditions! This is way too strong: there are infinitely many such possibilities.

But, though flawed, this latter account, and Salmon's work, relates to what is arguably the new received view of explanation: mechanical accounts of explanation. Indeed, this is threatening to morph into a latter-day version of the global philosophical scheme created by the positivists: it aims to provide accounts not only of explanation, but also reduction, discovery, laws, and more. The idea is simple: to provide an explanation is to provide an account of the mechanism that brought it about. This move to mechanisms coincided with a move away from the focus on physics by philosophers of science, with greater attention paid to biology and the life sciences. In such cases the logical accounts look wholly inappropriate, while mechanical accounts seem rather natural. To link up to the topic of the next chapter, we might say that the search for mechanisms also offers a pretty good characterization of how science operates, and where pseudosciences fall short. Of course, as you might have guessed, the problem is now to make sense of the notion of mechanism itself in a way that does not fall prey to problems of induction, laws, and causality.

Summary of Key Points of Chapter 2

- Logic plays a key role in philosophy of science, forming the basis of once common accounts of scientific reasoning, confirmation, and explanation. A key distinction is between inductive and deductive reasoning. While the latter leads to certainties, the former can at best offer *support* to some idea or theory. Yet many accounts of science state that scientific reasoning involves inductive reasoning.
- Several serious problems and paradoxes arise from this inductive foundation, infecting the deepest levels of our knowledge of the world. The most serious of these is Hume's problem of induction, which denies the very possibility of justifying inductive reasoning, and thereby scientific reasoning if it is so understood.
- An empiricist account of the laws of nature faces

similar problems too, on account of not having the requisite strength to cover unobserved (or future) scenarios: they fail to ground the kind of necessary connection we associate with laws of nature. Yet alternatives face problems in making sense of that very necessity.

- Such problems of logic and induction go on to infect accounts of explanation too, since one once orthodox view demands that any and all scientific explanations must contain at least one law of nature. The presence of such laws is required to ground the fact that some feature in need of explanation follows logically from the premises of a deductive argument. The over-employment of logic leads to many counterexamples, pointing to the need for an alternative account.

Further Readings

Books

- Many of the examples in this chapter come from the superb collection of essays in M. Salmon et al. (eds.), *Introduction to the Philosophy of Science* (Hackett Publishing Company, 1999). This is a collection of introductory essays written by a stellar team of then-members of the Department of History and Philosophy of Science at the University of Pittsburgh.
- By far the best book-length treatment of the problem of induction is Colin Howson's *Hume's Problem: Induction and the Justification of Belief* (Oxford University Press, 2000).
- A beginner's guide to causal inference and its problems is: Judea Pearl and Dana MacKenzie, *The Book of Why: The New Science of Cause and Effect* (Basic Books, 2018).
- The original statement of the problem of induction (and causation) can be found in David Hume's *A*

Treatise of Human Nature (Oxford University Press, 1739) – still a very good read!
- Nelson Goodman's grue example can be found in his book, *Fact, Fiction and Forecast* (Harvard University Press, 1955).

Articles

- An excellent study of Sherlock Holmes' methods from a philosophy of science perspective is: L. J. Snyder, "Sherlock Holmes: Scientific Detective." *Endeavour* (2004) **28**(3): 104–8. Snyder in fact argues that Holmes' methods follow Bacon's principles.
- A brief argument concerning the power of Hume's problem of induction is Colin Howson's "No Answer to Hume." *International Studies in the Philosophy of Science* (2011) **25**(3): 279–84.
- A good brief review of the problems facing accounts of laws of nature can be found in Fred Dretske's "Laws of Nature." *Philosophy of Science* (1977) **44**(2): 248–68.

Online Resources

There are several excellent entries from *The Stanford Encyclopedia of Philosophy* on topics relating to the present chapter:

- Leah Henderson's "The Problem of Induction," *The Stanford Encyclopedia of Philosophy* (Fall 2019 Edition), E. N. Zalta (ed.): plato.stanford.edu/archives/fall2019/entries/induction-problem.
- James Woodward's "Scientific Explanation," *The Stanford Encyclopedia of Philosophy* (Fall 2017 Edition), E. N. Zalta (ed.): plato.stanford.edu/archives/fall2017/entries/scientific-explanation.

- Vincenzo Crupi's "Confirmation," *The Stanford Encyclopedia of Philosophy* (Winter 2016 Edition), E. N. Zalta (ed.): plato.stanford.edu/archives/win2016/entries/confirmation.
- John Carroll's "Laws of Nature," *The Stanford Encyclopedia of Philosophy* (Fall 2016 Edition), E. N. Zalta (ed.): plato.stanford.edu/archives/fall2016/entries/laws-of-nature.

BBC Radio's *In Our Time* program has several excellent episodes on relevant topics:

- Hume, in which Hume scholars Peter Millikan and Helen Beebee and others cover a range of topics including induction: bbc.co.uk/programmes/b015cpfp.
- Laws of Nature, including Nancy Cartwright: https://bbc.co.uk/programmes/p00546x5.
- The Scientific Method, with Michela Massimi and John Worrall: https://www.bbc.co.uk/programmes/b01b1ljm.
- Baconian Science, including Patricia Fara and Stephen Pumfrey: bbc.co.uk/programmes/b00773y4.

Bryan Magee's 1978 BBC TV series *Men of Ideas* contains an excellent discussion of Hume's problem of induction, with John Passmore: youtube.com/watch?v=UJLHf9Vt-m4.

3

Demarcation and the Scientific Method

Is science *special*? If so, what is it that makes it so? In the early days of philosophy of science, one of the main goals was to identify rules governing what makes science scientific, in order to distinguish science from non-science and, importantly, pseudosciences masquerading as the real thing. One of the major questions that the logical positivists focused on was what exactly gets to have this special title of "science" bestowed upon it, and how can certain subjects such as the social sciences be made more scientific? We still face this issue all the time in everyday life: is homeopathy scientific? What about climate science? String theory? Who gets to decide? This is known as the problem of demarcation.

This chapter presents a general overview of the problem and proposed solutions, but the main focus is on a case study in which the demarcation problem became a legal matter concerning the teaching of "creation science" in high schools. The philosophical aftermath of this episode spelled the demise of the notion of a demarcation problem in the original sense of supplying a set of conditions or "demarcation criteria" that one could tick off to check for "scientificity," and this is probably still the most commonly held view amongst today's philosophers of science. Science remains the best tool we have for assessing the veracity of claims about the natural world. Yet the notion that there *is* something special about the way science operates, along with the idea that it seems to lead to reliable knowledge about the world, remains to be fully understood. With the levels of trust in science so low at present, it would be no bad thing if this topic became once again the central locus for philosophy of science.

From Verification to Falsification

The question of what distinguishes science from other human endeavors is known as the "demarcation problem." It is usual to separate out these endeavors, rather artificially, into three classes: "science," "pseudoscience," and "non-science." We are mainly interested in the distinction between science and pseudoscience, since there is often dispute between these two (i.e. pseudosciences are often paraded as *bona fide* sciences, thus encroaching on scientific territory), whereas there isn't usually a clash between science and non-science. Of course, the pseudosciences are strictly viewed as being contained in the class of non-sciences (by scientists), but they are a particularly naughty subclass in that they claim to be otherwise. Table 3.1 gives a few examples.

Table 3.1 *Examples of sciences, pseudosciences, and non-sciences*

Science	Pseudoscience	Non-science
General Relativity	Astrology	Poetry
Paleontology	Homeopathy	Religion
Evolutionary Biology	Intelligent Design	Metaphysics

The question is, then: what makes general relativity scientific and astrology not? Why is astrology just a pseudoscience? Though not all philosophers of science agree that we can make such simple divisions, we might think that it is important that we are able to for a variety of reasons. Of particular importance is to "protect" society from the spread of misleading ideas, ideas that might be harmful. For example, if the claims of creation scientists (to be discussed later in this chapter) are deemed just as scientific as evolutionary theory, this would allow religious leaders to evade the common separation of state and church in Western society.

The traditional response to the demarcation problem is to say that it is a particular type of *method* that so demarcates. Most often, it is an inductive methodology that distinguishes science from non-science. The logical positivists really started the whole business of demarcation, and the search for a demarcation criterion: they said that scientific statements were all and only those that can be *verified* by observation. Those that cannot be so verified were deemed "meaningless" or "metaphysical." Of course, this anti-metaphysical stance is related to the "positivistic" stance, and the distinction of science from pseudoscience was only a concern inasmuch as the latter overlapped with metaphysics too. That this is in fact the case can, however, be discerned from the fact that metaphysical claims tend to fit with all empirical evidence, and one of the signs of a pseudoscience is the same quality: fitting all possible evidence and therefore violating testability. Yet this idea faces numerous problems, as we saw in the previous chapter: for now, we can just repeat the broad ideas and refer the reader to that chapter for details. Induction has no justified basis: one has to use induction to defend it ("Hume's problem"). Moreover, the notion of verification is logically problematic since one simply could never verify a theory (intended to be general) from observations. For example, given the standard way of writing laws (see section "Laws of Nature" in chapter 2), $\forall x Fx \supset Gx$, no number of things x that are both F and G (e.g. metals and expand when heated) will be sufficient to verify the theory with certainty. Observation too, involved in the verifying of the theory according to empiricist principles, has its own problems: it is hard to set up any notion of observation that doesn't involve theoretical terms already (we turn to this problem in the next chapter). Because of these, and related problems, Karl Popper dispensed with induction and the primacy of observation, and so side-stepped all these problems. This side-stepping offered a new response to the demarcation problem, and indeed Popper viewed this as the central problem for philosophy of science.

Popper's answer to the problem involved providing a deductive criterion to distinguish between science and non-science (and pseudoscience, which was his primary target). This is known as the "falsifiability criterion": science is special – distinct from pseudoscience – because it is falsifiable. That is, scientific claims ("conjectures," which do not need to be generated inductively) are open to *refutation* by experimental evidence. The potential for *conflict* with observation is key, not harmony with it as the logical positivists thought. This negative approach constitutes the proper meaning of "testability" for Popper. To highlight this feature, Popper invoked the example of Einstein's prediction of the bending of light from a distant star around the Sun (see figure 3.1). This was a crucial experiment according to Popper, in that it was actually testable and would have refuted Einstein's theory had it not turned out to be the case. It was tested and confirmed in 1919 during a solar eclipse. In the process, Newton's theory was also refuted (though there were already problems with its treatment of the behavior of Mercury, which contributed to the creation of general relativity in the first place).

The idea is, then, that truly scientific claims are potentially *refutable*: that is what is special about science.

Figure 3.1 The photographic plate confirming Einstein's prediction that distant light would be bent around the Sun, which would be visible during a solar eclipse. Distant stars are slightly displaced from their usual positions by an amount predicted from general relativity. Popper was impressed by the fact that general relativity stuck out its neck at the risk of being refuted by this prediction
Source: Wikimedia Commons

Pseudoscience is not like this. Neither are metaphysical claims. Hence, the claim that God exists is not refutable; that is what distinguishes it from science. If a piece of evidence turns up that apparently conflicts with some picture of God, then we might be told that "God works in mysterious ways," for example. The claims of astrology and Freudian psychoanalysis are (according to Popper) also unfalsifiable, and so not scientific. What this means, to drill the point home, is that there is no piece of evidence we could have that would ever contradict the claims of astrology or psychoanalysis: any outcome can be made to fit. They transcend empirical evidence in this sense.

The logic of this idea is very simple, and involves a principle of inference known as *modus tollens* – its striking simplicity is probably responsible for its strong hold amongst scientists, though this same simplicity is what bothers philosophers. Let T be some theory or hypothesis which has as a (deductive) consequence the observation statement O – in other words, we can logically *derive* O from T. We can write this in symbols as: $T \supset O$ = "if T is the case, then O is the case too." What this implies is that if we find out that O isn't the case, then T can't be true either: $\neg O \supset \neg T$ = "if it is not the case that O, then T is not the case either" (the symbol "$\neg$" just means "it is not the case that" or just "not"). Why? Because T implies that O is true, so if we find out that O isn't true, T couldn't have been true after all.

Suppose T is some biological hypothesis, say: "scarcity of food causes cannibalism in primates." Then O might be something like: "there will be cannibalism among primates when there is little or no food." (This is very simple, but it should get the point across.) We now, as good scientists, go about testing the hypothesis. How do we do that? We find a situation where there are primates and little or no food (we may have to engineer this, or it may be that there is a natural experimental setup ready and waiting). If we find that there is indeed cannibalism then T (the theory that led us to make the particular observation) is, in Popper's terminology, *corroborated*. This does not mean that theory is *proven*, however, and there is a good logical reason for this, as covered in the previous chapter. The argument would look like this:

$$\begin{array}{r} T \supset O \\ O \\ \therefore \neg\Box\neg T \end{array}$$

Here the "$\Box$" symbol means "necessarily," so "$\neg\Box\neg T$" means "not necessarily not." You might think that we could infer the truth of the theory from an observation O. That would look like this:

$$\begin{array}{r} T \supset O \\ O \\ \therefore \mathrm{T} \end{array}$$

However, as we know from the previous chapter, that is a fallacy of logic known as *affirming the consequent* (in the expression $T \supset O$, T is the antecedent and O is the consequent). This explains the reason for Popper's claim that theories can at best be corroborated: they can survive empirical tests.

Now, back to our example with the primates. If we hadn't observed O (i.e. if we observed $\neg$ O, and so no cannibalistic behavior in a low food environment), then we would have the following logical steps:

$$\begin{array}{r} \mathrm{T} \supset \mathrm{O} \\ \neg \mathrm{O} \\ \therefore \neg T \end{array}$$

This is a valid argument. If the theory says O and we find that ¬O, then T must be wrong. This is the incredibly simple logic behind falsification.

To this day, falsificationism is the preferred stance of scientists. It is often invoked by anti-string theorists, who quickly point out that since string theory predicts almost *any* possibility, so that it can be made compatible with any observational or experimental results, it violates Popper's criterion and so is strictly unscientific. I recall hearing Richard Dawkins once describe in glowing tones how at a conference which had seen an older scientist's theory disproven, the scientist gracefully accepted that his theory had been demolished by this same simple logic. However, popular or not, this account fails for a number of reasons.

Firstly, let us take the rejection of induction. Hans Reichenbach considered the following "pragmatic problem": we base our technological innovations on the best available scientific theories. That is a perfectly rational thing to do. Yet we don't consider this as resting on the fact that these theories are so far *unrefuted*. That would surely be an *irrational* thing to do. We might use some laws from a theory of mechanics, that has not been refuted so far, but we won't build the bridge using this theory for this reason: we will use it because it has performed well *in the past*! This makes it reliable. That is, we reason *inductively* from past instances to future ones: we need to know that the bridge will be stable after many crossings! Popper wanted to have a purely *deductive* basis for scientific reasoning and progress (not based on inferences from past to future, or from particular instances to general cases), and so couldn't countenance such distasteful inductive ideas – we already noted in the previous chapter how Popper thought that his method also resolved "the problem of induction," one of the most serious problems in all of philosophy. All the worse for falsificationism. However, Popper had reasons to avoid induction in the present context, since if the scientific status of a theory is based on induction (on generalizing from observation and experience), then astrology might slip through as a science – and he didn't want that!

In terms of falsification itself, even before Popper came up with his falsificationist "demarcation criterion," in 1906, Pierre Duhem had already devised a problem for it. The problem concerns whether theories can even really be said to be falsified by *single* instances in the manner suggested by Popper. Logically it is beyond doubt, but what about as an account of real science? Duhem's argument involves the "holistic" idea that theories are never tested "in isolation," for there are always many other "auxiliary assumptions" that come into play. For example, we might consider Newton's theory of mechanics coupled with universal gravitation as providing a fairly tight theory that can make predictions and be tested independently of any other facts and assumptions. When we make a test, using a telescope to determine the position of some planet at some time in the future on the basis of Newton's theory, we may naively think that this is a solid test. Given this we might think, following Popper, that if we do not see the computed result through the telescope, then Newton's theory is wrong; it will have been refuted in an instant. But this is wrong, says Duhem. We *do* need additional assumptions, and ones that are highly theoretical. For example, we need to make assumptions about the propagation of light between the planet and our telescope; about the extent to which the light is refracted; about the interaction between the light and the telescope, etc.

Thus, around every theory is a "belt" of auxiliary hypotheses. If we were to get a result that went against the theory, we could, says Duhem, simply (or perhaps not so simply) make an adjustment in the belt, or remove one of the assumptions from the belt, leaving the core theory intact. Popper, then, viewed the situation as follows: we can deduce a number n of observational consequences C_i ($i = 1, ..., n$) from a theory T. If *any* one of these observational consequences is wrong (not observed) then the theory will be wrong too (because we *deduced* the Cs from T). Duhem says this isn't right: we deduce observational consequences C_i not from a single isolated theory T, but from the conjunction $T \wedge A_i$ of T with further auxiliary assumptions A_i. But then if we find that some observational consequence is wrong, we don't thereby have to conclude that T is wrong now, since it may also be some A_i that is wrong: we can't consider a theory to be refuted by some observation! There is something more to science and scientific progress than falsification alone. But note that sometimes an observation might be enough to reject a theory because our auxiliary hypotheses are *solid* (more solid than the theory).

We can put this in the language of logic, used earlier, as follows. The *real* argument is better represented by:

$$T \wedge A_1 \wedge \ldots \wedge A_n \supset O$$
$$\neg O$$
$$\therefore \neg(T \vee A_1 \vee \ldots \vee A_n)$$

In other words: O is not derived from the theory *T* alone, but from the theory *and* a host of auxiliary theories and hypotheses. So, what this means is that if we find that O is not true, then the theory is not refuted, because it could be any one from the list of auxiliary hypotheses that is refuted (the symbol "$\vee$" just means "or").

Indeed, Popper's favorite example of the deflection of starlight around the Sun was subject to exactly this kind of questioning in the aftermath of the experiment. In a 1930 paper entitled "The Deflection of Light as Observed at Total Solar Eclipses" (*Journal of the Optical Society of America* 20(4): 173–211), Charles Lane Poor noted that, not only did the deflection results not imply Einstein's theory of spacetime – since it only involved a retardation of light in the vicinity of large masses like the Sun – but also the results themselves were not without problems. There were no investigations studying the effects of temperature on the instruments, nor for checking the possible effects of abnormal atmospheric conditions during the eclipse. One could, in other words, point to some other factor as causally responsible, so that the theory is not confirmed. It is not a crucial experiment, in other words, and even if the result had been different, one could point to the same potential problems that Poor noted.

The Normal and the Revolutionary

In the previous chapters the focus was very much on the end-products of the scientific enterprise: the theories and models. It matters not how they were created; they might have just as well been dropped onto the Earth by aliens. Thomas Kuhn's book, *Structure of Scientific Revolutions*, signaled a "practical turn" in philosophy of science, in which the focus shifts to what scientists actually do, as opposed to idealized reconstructions of philosophers, such as the "statement + logic" views of the positivists and Popper.

According to Kuhn, Popper had not understood properly how science actually works: to do this one needs to pay attention to the history of science and look at how scientists actually operated. Kuhn finds that activity in science takes two forms: "normal" and "revolutionary." Normal science accepts a theory as true, and ignores foundational issues, content instead with "puzzle-solving." Failure to solve a puzzle is a failure of the scientist rather than the theory which, at this stage, is concrete. However, if there are repeated failures then this might signal a period of revolutionary science or crisis in which foundational issues *are* considered, triggered by some anomaly that will not go away. This will inevitably lead to the construction of a successor theory or a new paradigm. Only then is the old theory considered *refuted*, and we have a scientific revolution (some will, of course, be more significant than others). Hence, Popper's idea that scientists go about trying to refute theories at every opportunity simply does not fit the true historical picture, according to Kuhn, and really it would be a chaotic scenario if it did. Instead, Popper's scheme fits one quite rare part of the development of science which is reached as a theory is in extreme crisis and the entire theoretical edifice lies in the balance.

Though there are several important points of agreement between Kuhn and Popper – for example the rejection of the logical positivists' view of scientific progress via the steady accumulation of knowledge in favor of the replacement model based on revolutions – Popper was not at all impressed with Kuhn's normal science idea, which clashed with his political nature: science was the very model of a critical mindset; there should be no let up on the testing and questioning of hypotheses. Normal science meant not questioning the status quo. Yet Kuhn was able to fashion his own demarcation criterion based on just such normal science (rather than the revolutionary stage), in direct opposition to Popper's approach. Kuhn's answer to why astrology is a pseudoscience fits his vision of how science works: astrology was dropped after numerous failed predictions, much unreliability, and once something better had come along (with the Copernican revolution). The explanation for why it is, and was, never a genuine scientific theory is because there was never the "puzzle-solving" tradition nor a core theory to speak of; there were simply many rules of thumb. In this respect it is more a *craft* than a science.

Lakatosian Research Programs

Like Duhem and Kuhn, the Hungarian philosopher of science Imré Lakatos did not view theories as isolated, logical entities. Rather, he viewed theories as having three components: a "hard core," a "protective belt," and a "positive heuristic" (see figure 3.2). The hard core consists of the central laws of the theory (Newton's laws of motion, for example); the protective belt consists of Duhem-style auxiliary assumptions; and the positive heuristic is a part telling scientists how to respond to potential problems and anomalies in the theory by revising the protective belt, much in the manner of Charles Lane Poor's attempt to deny the orthodox interpretation of the starlight deflection experiment. Thus, theories are not static entities, but dynamically evolving processes, and they are rarely if ever falsified in the simple way Popper suggests. Scientists will often stubbornly stick with a theory even in the face of an apparent refutation.

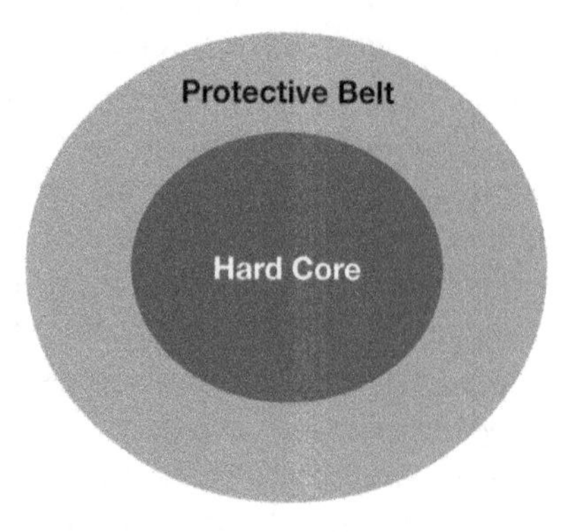

Figure 3.2 Theories according to Imré Lakatos: a hard core of fundamental results sits at the center, immune from revision, surrounded by a protective belt of auxiliary assumptions that, unlike the hard core, can be revised in the light of anomalies (the positive heuristic)

This is not irrational according to Lakatos, since it can often transpire that it leads to advances in science that would have otherwise been lost to an overly stringent method.

The example used by Lakatos is taken from Newton's celestial mechanics: Uranus was found not to move as Newton's theory (the hard core of it) predicted. But the theory wasn't abandoned, as Popper would have it and as the bare logic of the situation would seem to demand. Instead, a new planet was postulated in an area of the solar system that should then lead, given Newton's theory, to exactly the observed (apparent) discrepancy in Uranus' motion (this is the positive heuristic part acting on the protective belt) – a planet was indeed found: Neptune! The same set of procedures occurred again in problems with the observed motion of Mercury as compared with the motion as predicted by Newton's theory. Again, a new planet was postulated in some region (to be named Vulcan if it was found), but this time it wasn't found and a new theory (Einstein's new theory of gravity) was developed that better predicted the motion. But the replacement of Newton's theory had to await the further testing and development of Einstein's theory, which had to show itself to be more progressive as a research program, i.e. in terms of generating new predictions.

It is clear in this case that it was a good thing not to abandon Newton's theory on the basis of recalcitrant data because the data was not in fact a fault of the theory after all: the theory was functioning successfully, but the data to be plugged into Newton's equations of motion was incomplete. Even if it was a fault of the theory, the theory was producing excellent results elsewhere and there was no better alternative. Thus, Lakatos, like Duhem, thinks that theories cannot (and *should* not) be wiped out in an instant by "crucial experiments": it is a matter of how well a theory is doing relative to the field of theories. Yet falsification was Lakatos' preferred method of progress; just not at the breakneck speed Popper suggested and not without replacements waiting in the wings. In practice, scientists rarely, if ever, follow Popper's rules, at least not at pivotal moments in science.

Lakatos' positive suggestion involves an idea similar to Thomas Kuhn's, but it differs in an important way: whereas Kuhn says that theory change is largely irrational (a kind of "mob psychology" according to Lakatos), Lakatos claims that science works through competition between rival research programs (and is a *rational* process). Scientists will shift their allegiance according to how well a research program is doing: if one is degenerating while another is progressing, then rationally one will shift to the progressive one. If a research program *is* doing well, then a piece of evidence against the theory might not be enough to bother it. The problem with this is what to do in cases where there are no alternatives, or when a research program goes through a fallow period, but might nonetheless be able to progress later, perhaps as a result of some new technology.

In terms of the demarcation problem, Lakatos rather tells us what *doesn't* distinguish between science and pseudoscience:

- The number (or type?) of people who believe in a theory, and the strength of this belief.
- The claim that scientific theories can be *proven* from the observational facts.

Clearly there have been some bad pseudoscientific theories in history that have commanded impressive support (Newton's taste for alchemy, astrology, etc.). The second claim is simply an anti-inductivist one shared by Popper. However, Lakatos was radically against Popper's simplistic logical criterion on the grounds that it did away with the role of evidence. For Lakatos, the unit of demarcation was not statements linked by logical relations, but entire research programs, which can be progressive or degenerating. A progressive research program solves problems and adds to the store of empirical knowledge about the world. This condition characterizes science and amounts to a criterion for Lakatos: research programs that do not operate in this way amount to pseudoscience.

In a later section, in conjunction with a discussion about the battle between creation science (revised into the theory of Intelligent Design) and the Theory of Evolution by Natural Selection, we see how other attempts to solve the problem of demarcation try to build up a set of necessary and sufficient conditions: a set of conditions that a theory must meet in order to qualify as scientific. If just one of the conditions is not satisfied, then the theory is rendered unscientific (this is the meaning of *necessary* here); if all of the conditions are satisfied the theory is rendered scientific (this is the meaning of *sufficient* here). This totally ignores the kind of lessons from Kuhn and Lakatos that we see here and propagates an inaccurate model of science.

There is no Method!

The anarchists amongst you will be pleased to encounter the philosopher Paul Feyerabend (1924–1994). Whilst a native of Vienna, he was about as opposed to the Vienna Circle as it's possible to be. He considered their approach, and many of the other "sanitized" versions of science, to be almost morally wrong since it stunts real progress by placing limitations on what scientists can do. Method is, according to Feyerabend, nothing more than *rhetoric*: there are no real underlying principles regulating science. Rather, anything goes. One sometimes hears of people referring to religious texts as "works of fiction" or "fairytales" (or delusions, if you're Richard Dawkins). Feyerabend felt the same way about science, and indeed about *all* ideologies of which science was just another example:

> I want to defend society and its inhabitants from all ideologies, science included. All ideologies must be seen in perspective. One must not take them too seriously. One must read them like fairytales which have lots of interesting things to say but which also contain wicked lies, or ethical prescriptions which may be useful rules of thumb but which are deadly when followed to the letter (Feyerabend, "How to Defend Society Against Science" – from a talk given to the Philosophy Society at Sussex University in November 1974).

He views it as ironic that once science led the battle against dogma, superstition, and authority, but now seems to have landed itself in what it once fought against: power corrupts. Science claims to have found the path to truth. The argument involves the idea that science has found a *method* to get to the truth and there are *results* to prove it. We have already looked at this method, and suggested there are indeed problems with it. There are restrictions on Feyerabend's own anarchism, however:

it is a constrained anarchism! If objections are raised to some field, then the advocate must respond knowledgeably. They must consider the opponents' side, to see if their theory fits as many facts, and so on. The point is, there should be no unquestioned authority on what is. In terms of the demarcation problem, it is clear what Feyerabend thought: there simply is no such thing. Moreover, if science does indeed have some elevated status in society, it should not.

Indeed, as did Kuhn and Lakatos, Feyerabend looks to history for support – though he thoroughly rejects Kuhn's notion of normal science and thought Kuhn's famous book damaged philosophy of science by letting in too many bad thinkers with no knowledge of science itself. There are, he argues, many examples of past overlaps between what we would now think of as science of the highest order mixing with what we might now think of as lowlifes. Copernicus, he of the revolution, wrote on occult issues and consulted with numerologists and self-appointed magicians. If we look at the foundations of optics, it is infused with the work of craftspeople and artisans. Indeed, almost every advance in history has some element of the story that is utterly at odds with the rationalized picture of science we tend to be faced with today. Indeed, the way science is now set up means that even if one of the so-called pseudosciences *did* manage to show success, it would simply be absorbed as science after all, and would join the club.

Feyerabend didn't agree with the demarcation of science from other disciplines, but he did relish the opportunity to use the debate to force his ideas. On the other hand, Larry Laudan, in his "The Demise of the Demarcation Problem," argues that the whole demarcation problem debate is a bad one that we shouldn't be having. Indeed, the very terms of the debate, "pseudoscience" and "unscientific," should be removed from our vocabulary since they are essentially just emotive utterances, and hardly conducive to rational debate. Laudan's main point is that the whole business of demarcation does not differ in any important way from the distinction between "reliable" and "unreliable" knowledge. This is a debate that makes sense, and can be discussed without inflammatory rhetoric. Of course, the grounds for reliability still face Hume's problem. Indeed, we might view Laudan's evasion of the problem of demarcation as simply offering an old-fashioned inductive solution.

The Sciences of Creation and Design

The demarcation problem became a legal matter when, in the case of McLean v. Arkansas, a judge was asked to rule on whether "creation science" is a genuine science or not. More recently, this old battle between religion (or "the religious right") emerged once again, with creation science instead dressed up as "Intelligent Design Theory." Hence, this question isn't just pie-in-the-sky academic stuff considered only by philosophers who should be doing proper jobs! There is a practical dimension to the demarcation problem.

Creationism (or creation science) has its roots in a literal reading of the bible (Genesis in particular). It says that the claims contained in it can be separated off from religion and treated as scientific claims with scientific evidence to back it up. For example, the world was created in a sudden act by God; there was a catastrophic flood that should have left evidence in the geological record; the (species of) plants and animals that exist were made all at once independently at the origin of the world; the world was created just several thousand years ago, and so on. In 1995 and 1996 new laws were proposed in (the legislatures of) five US states that demand equal attention for evolutionary theory and creationism. There is some relevant precedent to consider here. In 1925 a law, the Butler Act, was passed

Figure 3.3 "A Venerable Orang-outang": a caricature of Charles Darwin as an ape, published in *The Hornet*, 1871
Source: Wikimedia Commons

in many southern states forbidding the teaching of evolution in schools: as they saw it, a religious person could not believe in evolution – Darwin agreed. One of these states was Tennessee, in which a school teacher (from Dayton), John Thomas Scopes, elected to be prosecuted for teaching evolution, to make the situation public. Scopes was found guilty: obviously, since he admitted to teaching evolution and there was a state law forbidding that. The trial became known as the "monkey trial," and generated a mild media swarm (see figure 3.3). It wasn't until the 1960s that the law was finally abolished (by the Supreme Court) on grounds of being unconstitutional (violating the separation of church and state).

Creation scientists believe that evolutionary theory is

wrong. In its place, they propose alternatives as given in Act 590 (from the Arkansas Annual Statutes of 1981):

> (a) "Creation-science" means the scientific evidences for creation and inferences from those scientific evidences. Creation-science includes the scientific evidences and related inferences that indicate: 1) Sudden creation of the Universe, energy and life from nothing, 2) The insufficiency of mutation and natural selection in bringing about development of all living kinds from a single organism, 3) Changes only within fixed limits of originally created kinds of plants and animals, 4) Separate ancestry for man and apes, 5) Explanation of the Earth's geology by catastrophism, including the occurrence of a world-wide flood, and 6) A relatively recent inception of the Earth and living kinds.

So: a main claim is that, though limited evolution can occur, it has to be restricted to kinds created originally. So an original finch was created, and this altered according to circumstance, but it could never become a non-finch; it can't change into another kind. In fact, part of the strategy of creation scientists was to also discredit the theory of evolution using something like demarcation criteria. Here's a passage from one of the main defenders of creation science, Duane Gish (all taken from "Creation, Evolution, and the Historical Evidence." *The American Biology Teacher*, 1973, p. 139):

> There is a world of difference, of course, between a working hypothesis and established scientific fact. If one's philosophic presuppositions lead him to accept evolution as his working hypothesis, he should restrict it to that use, rather than force it on others as an established fact.

The state of Arkansas passed Act 590, requiring "equal time" for evolutionary theory and creation science. However, there were objections made to the Federal courts pointing out that religion cannot be taught in schools. Creation science advocates claimed that it wasn't in fact a religious theory at all, and they classed it as a scientific theory. Calling in "expert witnesses," the presiding judge in the case instead argued that it wasn't science either. One of these witnesses was Michael Ruse, a philosopher of science (primarily of biology), who attempted to define a set of *demarcation criteria* (necessary conditions) for science. Larry Laudan objected to Ruse's strategy on the grounds that it is based on a myth about how science and scientists work, as we have already seen earlier.

Ruse's expert witness report states that, in his view, the first and most important characteristic of science is (1) that it relies exclusively on blind, undirected natural laws and naturalistic processes. And its claims must be (2) explanatory, (3) testable, (4) tentative, and (5) falsifiable. Creationism satisfies none of these, says Ruse. Judge Overton took up every single one of these conditions as "essential characteristics" of science and found creation science wanting.

Intelligent Design (ID) is really a *cosmological argument* for the existence of God. It states that "certain features of the universe and of living things are best explained by an intelligent creator, not an undirected process such as natural selection" (Discovery Institute website: https://www.discovery.org/id/faqs/). For example, it says that certain microscopic structures of biological organisms – such as the *flagella* ("outboard motors") of bacteria – are "irreducibly complex," meaning that the removal of any one part would result in the malfunctioning of the organism (see, if you really must, Michael Behe's *Darwin's Black Box: The Biochemical Challenge to Evolution*, Simon & Schuster, 1996, for this argument). The argument for an intelligent designer then claims that evolution, or any stepwise process, could not be responsible for features of this kind: hence, an intelligent designer must be responsible instead. ID theory is that it isn't incompatible with evolutionary theory *per se*: only certain parts of evolutionary theory are at odds with ID, and these are precisely the irreducibly complex bits. ID theory posits the existence of an intelligent designer whose interventions guide the process of evolution in such a way that irreducible complexity can arise. This intelligent designer is not specified, and this is how the ID theorists hope to get around the constitution forbidding the teaching of religion in schools. Suggested possibilities for intelligent designers are aliens and time-travelers – the alien hypothesis is a belief of the Raëlians: https://www.rael.org! Both of these are clearly preposterous as fundamental explanations since they simply shift the explanatory task back a step: what about their irreducibly complex features? If the same explanation is given for this, invoking yet more aliens and time-travelers, then an infinite regress threatens.

In 2005, in the case of Kitzmiller et al. v. Dover Area School District et al., the presiding judge, John E. Jones III, ruled that ID was *unscientific* and that it was "unconstitutional" for a pro-ID disclaimer (saying that "[ID] is an explanation of the origin of life that differs from Darwin's view") to be read out to students taking evolutionary theory courses in public schools. Judge Jones came up with a set of criteria, satisfaction of which makes a subject matter scientific. This harks back to the earlier McLean v. Arkansas trial, of course. However, they are not so much criteria, as suggestions for criteria:

- ID invokes and permits supernatural causation (thus violating the, apparently, "centuries-old ground rules of science").
- ID falls prey to the troubles that creation science fell prey to in the 1980s (concerning irreducible complexity).
- ID's negative attacks on evolutionary theory have been refuted by "the scientific community."

The first remark involves "methodological naturalism"; it states that the *supernatural* should be rejected by science as a matter of principle. What drives ID? Life, and the origin of life, is extremely statistically improbable: it involves states of high complexity, and high complexity equals low probability. ID theorists think that it is so improbable that no good naturalistic explanation can be given. Evolutionists say that evolutionary theory can cope just fine, and it is a lack of imagination or brain-power that prevents ID theorists seeing that this is so. This is why they deny naturalism.

It is certainly true that science operates according to methodological naturalism, but whether it *should* (always) is another matter. Hence, it may well be descriptively accurate, but it is not necessarily prescriptively accurate. We can easily imagine scenarios whereby naturalism conflicts with the evidence, where some supernatural beings really are trying to intervene in the world and communicate perhaps. In this case, if science were to stick by its maxim of methodological naturalism, then it would not be interested in the truth; yet truth is supposed to be an aim of science (according to many, at least).

Robert Pennock is a philosopher of science. Like Ruse, he was an expert witness at the trial. Like Jones, he focuses on the exclusion of supernatural elements from science. Supernatural hypotheses, they say, are not testable. This is surely wrong: breaks in the natural order of things should be very easy to test! Hypotheses about ghosts and telepathy *are* clearly testable. One can set up randomization experiments to see if someone is genuinely telepathic – indeed, early randomization ideas originated in such testing. One can imagine all sorts of possible supernatural phenomena that could be empirically tested in ways we would intuitively suppose are good scientific methods. Jones even writes that "while ID arguments may be true, a proposition on which the court takes no position, ID is not science" (p. 64). That is silly nonsense: it renders science a bizarre enterprise. One should not stake the future of science on a bet that involves naturalism as being the only way to go. One should allow that new evidence could knock naturalism off its pedestal.

Science doesn't always have to be true to count as science: Newtonian physics was the glory of science for hundreds of years, but it is false. That does not mean it wasn't science after all, and the scientists of the day were mistaken: it is simply a *false scientific theory*. The judges in these cases, and many scientists, are hellbent on proving that creationism or ID are not genuinely scientific. To do this, of course, they need a set of necessary and sufficient conditions on what science is, and when something counts as scientific. A better strategy would simply be to accept that the claims of ID are scientific – or perhaps follow Laudan and drop this way of speaking entirely – and then to simply show that the evidence doesn't support the claims, or, to go down a more Lakatosian line, that there are simply better explanations of the evidence.

But now suppose we accept that ID is simply false as a scientific theory. This leaves a tricky problem: Newtonian physics is false and yet is still taught in schools and universities. Is ID *more* false than Newtonian physics? Truth and falsity might seem like an "all or nothing" affair. However, Newtonian physics is, in a rigorous sense, a limiting case of theories we now believe to be correct: if we neglect high energies and small distances, Newtonian physics is a useful theory (it was used to get a man on the moon!). In other words, Newtonian physics gets *something* right: it predicts some things correctly, but fails to account for other phenomena that are anomalous according to it – we will see this point again in the final chapter. It is also useful to get to grips with the harder theories since, for example, quantum mechanics is based on Newtonian physics in certain deep ways. It is, moreover, historically important. Now, does it make sense to say ID gets something right? Or that it has a useful role to play in understanding some other established theory, or is a limiting case of some other theory, or that it is historically important? There is limited time for teaching, and much important material in the world, so common sense should prevail in restricting that time to those subjects that have a track record.

Summary of Key Points of Chapter 3

- A key problem for philosophy of science is how to distinguish science from other areas of inquiry and, especially, pseudosciences masquerading as scientific subjects. This is the problem of demarcation.
- Early responses focused on specific methods employed, in particular an inductive methodology (verificationism, in which claims of a scientific subject must be testable against observation or experiment) or a deductive methodology (falsificationism, in which claims of a scientific theory must allow for potential conflict with observation or experiment).
- The problem of demarcation becomes a real-world matter when it interacts with matters of public policy. Famous cases involve the teaching of creation science and Intelligent Design in public schools as scientific subjects. Such cases reveal flaws in overly simplistic logical responses to the problem of demarcation.
- One major reaction to such logical accounts was to turn to historical accounts, as followed by Thomas Kuhn in his book *The Structure of Scientific Revolutions*. This characterizes a science as something with a certain pattern of activity, known as a "paradigm." Other approaches followed along similar lines and still others (e.g. Paul Feyerabend's anarchism) denied any special epistemic status to science at all.

Further Readings

Books

- There are several excellent texts devoted to the demarcation problem, the most recent (and I think best) of which is the collection of essays edited by Massimo Pigliucci and Maarten Boudry: *Philosophy of Pseudoscience: Reconsidering the Demarcation Problem* (University of Chicago Press, 2013).
- The expert witness and philosopher of science Michael Ruse has a nice collection of essays targeting the case of creation science: *But is it Science? The Philosophical Question in the Creation/Evolution Controversy* (Prometheus, 1988).
- The best philosophical examination of Intelligent Design is an edited collection by Robert Pennock (the philosopher-expert witness in the ID court case): *Intelligent Design Creationism and Its Critics: Philosophical, Theological, and Scientific Perspectives* (MIT Press, 2002).
- Philip Kitcher has a wonderful study of the problems involved in creation science from a philosophy of science point of view in his *Abusing Science: The Case Against Creationism* (MIT Press, 1982).
- The documents from the original "monkey trial" can be found in: A. Horvath et al. (eds.), *The Transcript of the Scopes Monkey Trial: Complete and Unabridged* (Suzeteo Enterprises, 2018).
- Karl Popper's views are very nicely expressed, in their historical context, in his autobiography *Unended Quest: An Intellectual Autobiography* (Routledge, 2002).
- Thomas Kuhn's views are laid out in what is one of the most influential books of the last century: *The Structure of Scientific Revolutions* (University of Chicago Press, 1996).

- A superb collection of very relevant material, including correspondence between Lakatos and Feyerabend, can be found in: *For and Against Method*, edited by Matteo Motterlini (University of Chicago Press, 1999).

Articles

- A nice argument against the way the ID case was treated is Maarten Boudry, Stefaan Blancke, and Johan Braeckman's "How Not to Attack Intelligent Design Creationism: Philosophical Misconceptions about Methodological Naturalism." *Foundations of Science* (2010) **153**: 227–44.
- Larry Laudan's rejection of the problem itself can be found in: "The Demise of the Demarcation Problem," which is in R. S. Cohen and L. Laudan (eds.), *Physics, Philosophy and Psychoanalysis* (pp. 111–27). Boston Studies in the Philosophy of Science, vol. 76 (Springer, 1983).

Online Resources

There are again some excellent entries from *The Stanford Encyclopedia of Philosophy* on topics relating to the present chapter:

- Michael Ruse's "Creationism," *The Stanford Encyclopedia of Philosophy* (Winter 2018 Edition), E. N. Zalta (ed.): plato.stanford.edu/archives/win2018/entries/creationism.
- Sven Ove Hansson's "Science and Pseudo-Science," *The Stanford Encyclopedia of Philosophy* (Summer 2017 Edition), E. N. Zalta (ed.): plato.stanford.edu/archives/sum2017/entries/pseudo-science.
- Stephen Thornton's "Karl Popper," *The Stanford*

Encyclopedia of Philosophy (Fall 2018 Edition), E. N. Zalta (ed.): plato.stanford.edu/archives/fall2018/entries/popper.

BBC Radio's *In Our Time* program again has several excellent episodes on relevant topics:

- Popper, including Nancy Cartwright and John Worrall: bbc.co.uk/programmes/b00773y4.
- Logical Positivism, including Nancy Cartwright and Thomas Uebel: bbc.co.uk/programmes/b00lbsj3.
- Another nice radio portrait of Popper is: https://www.youtube.com/watch?v=_5J3cne5WEU (produced by Alan Saunders for the Australian ABC National Science program).
- Lorraine Daston, a philosophically-minded historian, discusses Kuhn's impact in Episode 2 of the Canadian Broadcasting Corporation's excellent 24-part series *How To Think About Science*: http://www.cbc.ca/radio/ideas/how-to-think-about-science-part-2-1.464988.
- An interview with Paul Feyerabend, including his views on demarcation, can be found at: youtube.com/watch?v=kDwoGtPbO5w.
- Audio of Imré Lakatos speaking on the demarcation problem can be found at: richmedia.lse.ac.uk/philosophy/2002_LakatosScienceAndPseudoscience128.mp3 – originally broadcast on June 30, 1973 as Programme 11 of The Open University Arts Course A303, "Problems of Philosophy."
- A huge database of links to documents and articles relating to pseudoscience in the area of health (where it is often labeled "quackery") can be found at: www.quackwatch.org.
- The BBC Radio series, The Infinite Monkey Cage, features Ben Goldacre and others with Brian Cox for a discussion of pseudoscience, in "When Quantum Goes Woo": bbc.co.uk/programmes/b051ryq8.

4

The Nature of Scientific Theories

The question "what are scientific theories?" is one of those that scientists rarely distract themselves with. They may ask "what is *Darwin's* theory?" or "what is *quantum* theory?," but never "what is a *scientific* theory?" This, like the demarcation issue, is really the preserve of philosophers of science. There have been very few answers to this question. One view in particular – that of, no surprise here, the early logical positivists and logical empiricists (especially Rudolf Carnap, Carl Hempel, and Herbert Feigl) – reigned for many years, so much so that it was for a time called "the received view" (though is more commonly called "the syntactic view"). However, many problems were found with this view, and since then a new orthodoxy has been taken up, with a trend towards semantics (i.e. what the theories are *about*) rather than syntax (encoding the purely formal aspects) and also towards more pragmatic viewpoints (associated with the more historical viewpoints of Kuhn, Lakatos, and others). In this chapter we look at these two general views about the nature and structure of scientific theories: "the syntactic view" and "the semantic view." Later sections will then consider one of the most important questions about scientific theories: how and to what extent they represent reality. Naturally, the answer we give to the first problem (what is a theory?) to some extent will determine one's answer to the second problem (how theories represent).

The Once Received View

Logical positivism formed around the time of two great revolutions in physics: quantum theory and the theories of relativity. It also took place during great changes in logic and mathematics: Gottlob Frege had just laid the foundations for propositional and predicate logic, Bertrand Russell and Alfred North Whitehead completed their work on the foundations of mathematics, and David Hilbert's work reintroduced the notion of "axiomatization" into science (attempting to eradicate hidden assumptions). This precision fed into the stance adopted by the logical positivist movement. The positivists viewed physics as the "paradigmatic science," and held the view that physics was the best (most *reliable*) method for knowing the world. They also held that the language of mathematics and logic, because of its precision and absence of ambiguity, should be the language of philosophy of science too. Their goal was to emulate these great advances in physics, mathematics, and logic – indeed, many of them had themselves contributed to these great advances.

The claim that physics is the most reliable way of coming to know the world and is a heavily mathematical discipline is clearly central here: the (epistemological) task the positivists set themselves was to understand how science was grounded in observation and experiment. To do this they considered the question: What makes statements about the world meaningful? They answered this question in two parts: firstly, they thought that natural language, with its imprecision and ambiguity, could pose a problem here and so formulated the sentences of science as sentences of a system of logic (first-order predicate logic). Secondly, they developed a criterion that showed how these sentences (of a scientific theory) related to the world, thus providing an empirical theory of meaning. This latter aspect was couched in terms of a new distinction between "theoretical" and "observation" sentences. This required some way of telling which sentences were true of the world, so that they were not just pure mathematics. This problem led to the formulation of the "verification principle": the meaning of a sentence is given by the procedures that one uses to show whether it is true or false. If there were no such procedures, then the sentence is thereby deemed meaningless (or "non-cognitive"). This allowed the positivists to dispense with ethics, metaphysics, religion, and pseudosciences in one fell swoop. Since experiment and observation is what makes science reliable for the positivists (thus distinguishing it from other types of knowledge), they needed some way of bridging the gap between the theoretical sentences of the scientific theory (that don't have an immediate counterpart in our observations) and those sentences expressing observations and experimental results (which wear their meanings on their sleeves as it were). Formally, they needed another set of sentences, "bridge sentences" (or "reduction sentences" or "correspondence sentences"), to fix their meanings (see figure 4.1).

So now we come to the logical positivist and logical empiricist characterization of a scientific theory (it was given the name "the received view" by Hilary Putnam in his takedown of the view), though it is somewhat formal as you might expect: *theories are sets of sentences that can be put into an axiomatic structure so that all of their logical relations and deductions can be made explicit.* We have already seen this idea in action in chapter 2, for the most important of these sentences were *laws*. There are, recall, two kinds of law: universal and statistical.

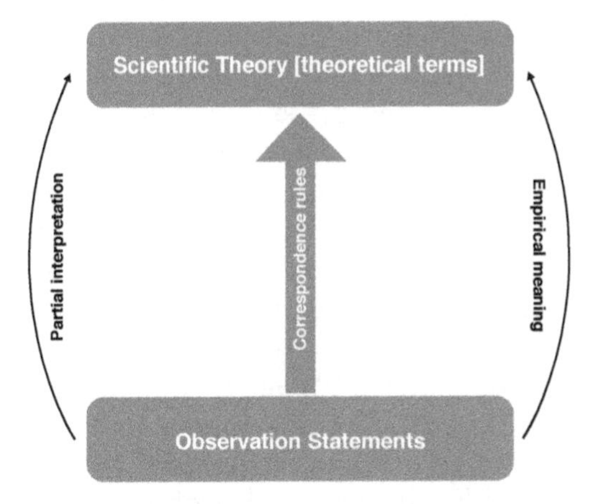

Figure 4.1 In the syntactic view of theories, the definition of scientific or theoretical terms from observation statements occurs through correspondence rules which provide the theory with real-world meaning

Universal laws have unrestricted application in space and time and have the logical form $\forall x \, F x \supset G x$ (or, "for all things x, if x has the property F, then it also has the property G"). Statistical laws, on the other hand, involve statements that make their conclusions more probable: x will be *more likely* to have G if it has F. These laws are used as an essential component in the logical positivist conception of scientific explanations: a particular observation sentence is deduced from a universal law (given some boundary or initial conditions – i.e. sentences describing the way the world is at a time). And, if the observation sentence was deduced prior to the corresponding fact being observed, then we have an instance of a *prediction*. If the theory gets the prediction right – i.e. if the observation sentence matches the observed facts – then the observation sentence gets *verified* and the theory from which it was deduced is *confirmed*.

According to a key member of this school (and the person that laid this idea of theories out most explicitly), Rudolf Carnap, there is a further way of dividing the laws in science: empirical laws and theoretical laws. The empirical laws are those that can be defined directly by empirical observations (they concern observable entities and their attributes). "Observable" here is a difficult notion: there are two senses. We have the philosopher's sense (corresponding to a more "commonsensical" notion), which simply refers to a property that is *directly* perceivable by the senses: e.g. "hard," "cold," "green," "rough," etc. And then there is the scientific sense that corresponds to a *measurable quantity* – "mass," "angular momentum," "energy," etc. The key difficulty here is that observation is not a simple concept: is it observation with the unaided senses or with a slight bit of aid, but not too much? It seems we have a continuum ranging from unaided perception to complex and indirect methods of observation (using spectacles, telescopes, microscopes, PET scans, etc.). It would seem that drawing a line between observable and unobservable is very arbitrary: our equipment is continually improving so that today's unobservables may become tomorrow's observables. Putting this aside, the empirical laws are those that have been attained by generalizing from particular observations and measurements. The theoretical laws contain terms referring to "unobservables." These are not arrived at by generalizing from observations and measurements (how could they be: they are unobservable!). This means that the theoretical laws are not fully justified by empirical facts. Theories are "partially interpreted calculi." The calculus is only "partially interpreted" in that only the "observation terms" are "directly [completely] interpreted" – the "theoretical terms" are only partially interpreted. Let us explain what all this means after a brief historical interlude.

Logical positivism (LP) and logical empiricism (LE) are often conflated; but there are significant differences. Both grounded epistemology (issues of warrant and justification for scientific knowledge claims) in empiricism (the view that observation, experience, and experiment is where knowledge comes from); both gave a central role to logic; and both rejected speculative metaphysics. However, logical empiricism is in part itself a reaction to logical positivism: there are fundamental philosophical differences. The two groups were more or less contemporaries (1920s and early 1930s), and sprang up in close geographical locations (LEs from Berlin and LPs from Vienna – the latter were also known as "the Vienna Circle"). The two groups were close early on, and their chief founding members (Rudolf Carnap from the LP camp and Hans Reichenbach from the LE camp) jointly founded and co-edited the journal for philosophy of science known as *Erkenntnis*. Political events drew an initial wedge between the movements: the rise of fascism caused the dispersion of the groups. Logical positivism faded by the second half of the twentieth century, and many from the Vienna camp changed their allegiance (including Carnap).

The chief problem with LP, as exposed by the logical empiricists, is that it is too restrictive in terms of what theories can succumb to its axiomatic approach – not only theories from biology and psychology were impossible to reconstruct in this format, but even central theories of physics. Moreover, it ends up ruling out much of what makes science what it is. For example, LP says that any scientific statement (theoretical, and hence dealing with unobservable features or not) has to be expressible in terms of observation. But observation reports are about past or immediate present experiences. Yet much of science is about the future: predictions are about the future. No scientific law does not involve claims about future events since the essence of a law is to "assure us that under certain given conditions, certain phenomena will occur" (Hans Reichenbach, "Logistic Empiricism in Germany and the Present State of its Problems." *Journal of Philosophy* (1936) 33(6),

p. 152). Hence, LP does not have the resources to deal with a massive chunk of science: in reducing scientific discourse to perceptual reports plus logic, we are stuck in the past and present. This is primarily what separated the logical empiricists from the positivists.

The central idea of logical empiricism is that the justification for all scientific knowledge comes from empirical evidence coupled with logic – included with "logic" here is induction, confirmation, and also mathematics and formal logic. However, unlike the positivists, this was not a phenomenological account: i.e. sense impressions (the stuff of immediate perception) are not what we observe; we observe middle-sized physical objects! Sense impressions are themselves the constructs of psychology.

The other issue concerns the reality of the unobservables (the stuff referred to by the theoretical terms). The logical positivists have no way of saying that the unobservable realm possesses any kind of reality; they can't infer the existence of genes from the fact that there are measurable, observable procedures we can use to talk about them. Reichenbach referred to this as a problem of *projection*. He gave the following example: imagine we were restricted for our entire lives (by some superior aliens conducting an experiment perhaps) to a cubical world with translucent walls; and by a complex series of mirrors and lights outside of the cube, shadows of birds are cast on the walls. An observer might well infer the existence of birds outside of the cube from the patterns of behavior exhibited by the shadows. Reichenbach says that a physicist makes just such inferences – the example of the gene might be better here, if we are talking about projection, for the structure of the gene was precisely inferred from a projection: x-rays are deflected from the DNA and cast an image on a photographic plate – and it is a legitimate inference to make on probabilistic grounds. But logical positivists, with their reductions to perceptual experience and logic, cannot make such inferences: the entire reality for them consists of the projection itself: the shadows and their patterns. There are no birds (and no genes). Inasmuch as such things can be referred to (using theoretical statements involving theoretical terms) they would have to say that they are the sense impressions alone, the shadows and the projection on the plate. (Reichenbach does not explain why he thinks this *is* a legitimate inference: however, he most likely had some kind of inductive logic involving "inference to the best explanation" in mind: the existence of *real* birds would render the shadows of the birds more probable, and that is the best explanation for the shadows.)

However, logical empiricism is a tricky position to pin down when it comes to its views on what theories are: its adherents seem to shift from the syntactic to the semantic account. What we can say is that the centerpiece of LE is its analysis of explanation, prediction, and confirmation, all of which are aligned to its more metaphysically open attitude (and its *realism* with respect to the unobservable/theoretical content of a theory). For example, logical positivism on the whole believed that the notion *scientific* explanation was meaningless: science deals with describing and predicting natural phenomena, and tries to systematize our knowledge of these phenomena, but answers to "why-questions" are no part of science but part of metaphysics or theology. However, the logical empiricists (especially Carl Hempel, who we met in chapter 2) thought that explanation was a highlight (if not *the* highlight) of modern science. Moreover, a fairly sharp explication of the concept could be given by philosophers (i.e. by him!).

All this is opposed to those who claim to follow Kuhn in arguing that matters of scientific theory depend on considerations outside of logic, and is a largely *irrational* affair – this wasn't Kuhn's stated position, but it is how he has been misinterpreted by many. This is a tricky point though: the logical empiricists were not claiming to describe how scientists constructed and chose theories, but offered instead *rational reconstructions* of the final products of science. Their vision of science was utopian. The idea, originally, was to use the best theories then available, relativity and quantum theory, to provide analyses of space, time, and causality – Einstein's analysis of simultaneity in terms of rods, clocks, and light signals, was the motivation behind much of this. What were broadly metaphysical notions were instead given a purely *scientific* elucidation in this scheme. Again, the view divides the non-logical (i.e. those other than "and," "or," "not," "all," "a," etc.) vocabulary of science into two parts:

- *Observation terms* – terms such as: "red," "big," "soft," "next to," etc.
- *Theoretical terms* – terms such as: "electron," "subconscious," "gene," "spacetime," etc.

The division of terms is based on the idea that observation terms apply to *publicly observable things* (and pick out observable qualities of these things), while the theoretical terms correspond to those things and qualities that aren't publicly observable.

The division of the terms in the scientific vocabulary is then taken to generate another division, this time dividing up scientific *statements*:

- *Observation statements* – statements containing only observation terms and logical vocabulary.
- *Theoretical statements* – statements containing theoretical terms.
- *Mixed statements* – statements containing a mixture of observation and theoretical terms.

Given this background, one can then define a theory as an axiomatic system which is *initially uninterpreted*, but which gets "empirical meaning" as a result of specifying the meanings of just the observation terms. A partial meaning is conferred onto the theoretical terms by this act ("by osmosis" is how Hilary Putnam puts it). Let's repeat the basic idea: a theory is viewed as an axiomatic deductive structure which is partially interpreted in terms of definitions called "correspondence rules" (this is the proper term for Putnam's "osmosis"). The correspondence rules define the theoretical terms that appear in the theory by reference to the observation terms. This still needs some unpacking.

First: what exactly is an axiomatic deductive system? This is just a structure that consists of a set of deductively related statements (sentences) – where the deductive relations making up the structure are provided by mathematical logic. In the case we are interested in, scientific theories, the statements (that are logically related) are generalizations (laws), some small subset of which are taken to be the *axioms* of the theory – axioms are unproven statements in the theory, used to specify the rest of the theory; it might be that the axioms specifying one theory are theorems (proven things) in another theory (e.g. as the laws of chemistry are axioms in chemistry, but theorems of atomic physics, which in turn has its own axioms). Within a theory, the axioms are the laws of the highest generality. All laws of the theory, with the sole exception of the axioms, can be derived from the axioms.

Axiomatization itself is a formal method that allows one to specify the (purely syntactic – i.e. no meanings) content of a theory. One lays out a set of axioms from which all else in the theory (all other laws, statements, etc.) is derived *deductively* as theorems. The theory is then identified with the set of axioms and all its deductive consequences (this is called the *closure* of the theory). An example is always best.

Let's consider the kinetic theory of gases. The structure of the kinetic theory of gases consists of (1) a set of laws (Newton's laws of motion) and (2) a set of singular existential and factual statements (telling us the actual conditions of the world). This set, with the laws and singular statements, is the "model of gas theory." It says various things about gases, such as: gases consist of molecules, molecules are minuscule, the number of molecules is huge, molecular collisions are perfectly elastic, molecules are in random motion, etc. Some of the terms here refer to unobservable entities and properties (molecules, random, etc.). So, the kinetic theory can do a great deal: it can explain the gas laws, laws about the rates of diffusion of one gas in another, and so on. Thus, theories *unify* diverse phenomena under the same set of laws. It also gives us a deep understanding of various concepts too: temperature, for example, is discovered to be the mean molecular kinetic energy – and we can define pressure and understand pressure in similar terms. Thus, theories can give us a *microscopic* structural account of common, macroscopic things.

But as it stands, the basis of a theory is a logico-mathematical construct. What has this mathematical monstrosity got to do with the empirical world of science? This is where the correspondence rules come in. The correspondence rules are definitions that link up the theoretical terms with observations, thus giving the theoretical statements empirical meaning (this has to be indirect, of course, because the theoretical terms refer to unobservable parts). So, let's focus on biology. There we have lots of theoretical terms: population, disease, fertility, gene, etc. These are unobservable. However, they are given meaning through observation: fertile, for example, is "partially" defined (i.e. indirectly) through reference to the outcomes of certain sexual events under certain specified conditions. It might be that some theoretical term is defined (gets its meaning) by another theoretical term, but when this happens the buck stops at some observation statement. This is the trenchant empiricism in action.

The account of theory construction in the syntac-

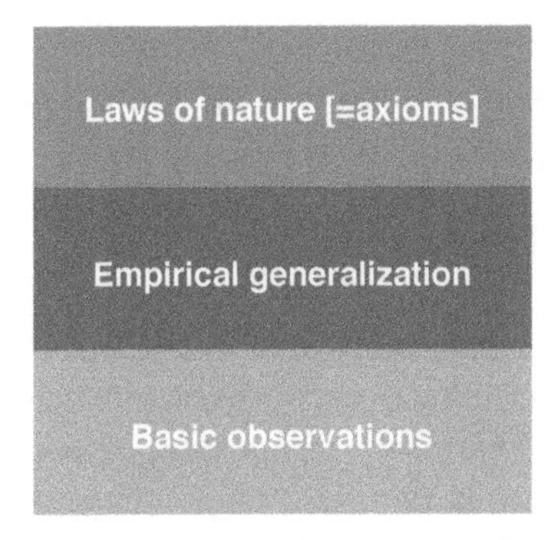

Figure 4.2 The layer cake view of the construction of scientific theories, in which progressive generalizations are made from a basic set of empirical observations

tic view can be described as a "layer cake" account. One begins by making inferences from particular observations (inductive generalization) to empirical generalization (constructed from the observation statements $\mathcal{O}$), and then from these empirical generalizations (via hypothetico-deductive inference) to laws of nature (constructed from the theoretical statements $\mathcal{T}$). Hence, the theory is built up in a series of layers starting with particular observations and ending up with laws of nature (see figure 4.2).

The tightly constructed logical nature of theories in the syntactic view means that theory change happens whenever any single element of the theory is modified, deleted, or added. The approach thus gives too fine an account of theories. Moreover, it does not allow one to view them as temporally extended entities with dynamics of their own. On this account they have no history, no gradual notion of theory construction or change. This is a serious problem for the syntactic view: we surely want a view whereby slight changes in a theory do not result in an entirely new theory. Yet this is not possible if they are defined as sets of sentences. Axiomatization seems to go against what the history of science tells us: many successful theories were constructed without a firm axiomatic basis. This did not make them non-viable.

Theory and Observation

So the whole setup above requires the existence of a split in the non-logical vocabulary of a scientific theory between observational terms and theoretical terms. This then grounds the distinction between observation statements and theoretical statements. But is either distinction really viable? The logical positivists never actually defined what they meant by "observable" and "unobservable," but instead gave examples: *electrons* are unobservable but *tracks* left by electrons in bubble-chambers are observable. However, they did make various assumptions about the distinction. The first assumption is that though there may be borderline cases in which we cannot tell if we have an observable or an unobservable, there are clear-cut cases too (electrons and their tracks) – hence, there is a clear distinction to be drawn (see figure 4.3, in which an anti-electron, or "positron," was discovered from its tracks). The second assumption was that the distinction was "theory-neutral"; that is, what is considered to be observable and unobservable does not vary depending upon what theory one holds about the world. Thirdly, the distinction is "context-neutral": it doesn't vary depending upon what questions scientists might be asking. Fourthly, the distinction would be drawn in the same way by both scientists and philosophers alike (i.e. regardless of their persuasions, whether realist or anti-realist). Fifthly, the distinction is based on a specific vocabulary associated with the observable and unobservable realms: what is observable is described by a special "observational vocabulary" and what cannot be observed is described by a special "theoretical vocabulary" (and never the twain shall meet!).

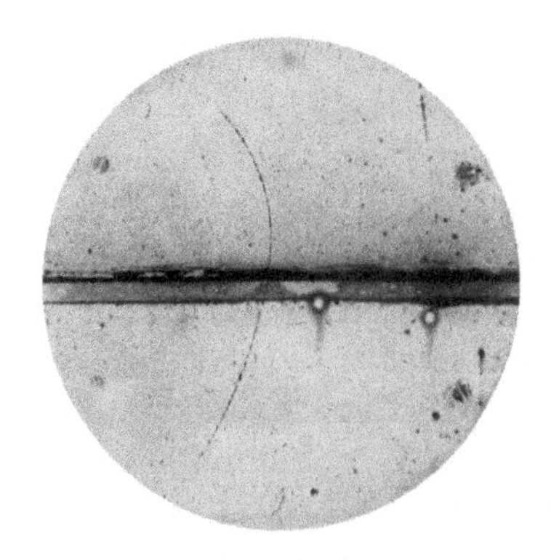

Figure 4.3 The first ever "observation" of a positron, by Carl Anderson, on August 2, 1932, using a cloud chamber to reveal the particle's trajectory

Source: Carl D. Anderson, "The Positive Electron." *Physical Review* (1933) **43**(6): 491–4

These assumptions can each be challenged. The second assumption has been subject to the best-known attack – by Norwood Hanson, Paul Feyerabend, and Thomas Kuhn, amongst others. The objection involves the claim that observation is "theory-laden." In other words, what you see (i.e. directly observe) depends crucially on what theories you happen to hold. When Aristotle and Copernicus looked at the Sun, they saw, quite literally, different objects: Aristotle saw a body that moved around the Earth, and Copernicus saw an object at rest around which the Earth and the other planets revolve. (A rather extreme, and so controversial, version of this idea was floated by a linguist, Benjamin Whorf, under the banner of "linguistic relativity." The basic idea is that language, like theory, strongly influences the kinds of conceptual experiences we are able to have. Whorf was led to his view after an analysis of the linguistic structures of various non-Western cultures, such as the Mayans, Aztecs, and Hopi. The idea is that one's very worldview is determined, to a large extent, by the linguistic conventions of the culture you find yourself embedded in. What this means is that there are thoughts that, say, a Hopi Indian can entertain that I, non-spiritual Westerner that I am, simply cannot understand. Again, our worldviews are, on this view, incommensurable.)

There are replies to this objection. The most famous is due to Fred Dretske, who argued that we need to draw a distinction between "epistemic" and "non-epistemic" seeing (observing). This amounts to the distinction between "seeing *that* x is F" and "seeing x." The former type of seeing is indeed theory-laden (it requires conceptual information), but the latter is not. Hence, Aristotle and Copernicus *saw* the same object, but saw it *as* a different thing due to their differing beliefs and theories.

The first and third assumptions have been challenged on the grounds that we can observe some theoretical entity through its effects, even though it itself is "hidden" from our senses. For example, a park ranger might see that there is a distant fire just by the smoke that is coming from it. An astronomer can observe a distant star when he sees a reflection in a telescope. Following this line of thought, it follows that electrons are observable since we can see the tracks in a bubble-chamber. So all of the things that positivists stuck in the bin marked "unobservable" are really observable after all. Also, since science continually invents new ways of observing the world, the second assumption takes a hit again: scientists tell us we can now observe black holes, which was once thought to be well-nigh impossible. More simply, consider the terms "spherical" or "dark." These would seem to belong to the category of observation terms. We can easily observe that a soccer ball is spherical, but what of an invisible speck of sand: is it spherical? We can observe that a room is dark, but what about the far side of the moon: is it dark? Is the application of the predicates (expressing properties of things, such as dark or spherical) in these latter cases observational? Such problems spelled the end for the syntactic view, and for the general framework for theories that generated it.

The Semantic View

There were two new directions the understanding of scientific theories went in: (1) a *historical* direction, which complained about the overly formal nature of the syntactic view, and (2) a *semantic* direction, which didn't have a problem with the formal nature of the syntactic view, but complained that it simply used the wrong formal machinery: sets of sentences rather than abstract models.

The semantic conception of theories began in the 1940s in the Netherlands, with the philosopher-logician Evert Beth. Beth's work languished somewhat until another Dutchman, Bas van Fraassen, developed it into a complete philosophical account of theories in the 1970s. The approach drew inspiration from both logic (in particular Alfred Tarski's work on formal semantics) and from physics (with John von Neumann's work on the foundations of quantum mechanics). The basic difference between the syntactic view and the semantic view can be seen with a simple geometrical theory, specified by the following three axioms (here employing an example due to van Fraassen):

- A1 For any two lines, at most one point lies on both
- A2 For any two points, exactly one line lies on both
- A3 On every line there are at least two points

Firstly, on the syntactic view, one would have to reconstruct these axioms in some appropriate formal language (involving quantifiers and the other logical machinery). Then one would have to introduce correspondence rules linking the theoretical terms (lines and points) to observations. The crucial difference, however, is in how the axioms are approached. The syntactic view looks at what can be derived from them: this notion of *deduction* is a purely syntactic feature. In the case of the semantic approach, the concern is with "satisfaction": *satisfaction* of the axioms *by* something. This is a purely semantic notion: the "somethings" that satisfy the axioms are known as "models" of the axioms. So the focus is not on the axioms as such (on non-abstract, linguistic entities) but on models (abstract, non-linguistic entities).

So, going back to the axiom-system above. We see that a possible model would be a single line with two points lying on it: A1 is trivially satisfied, since it talks about pairs of lines. A2 is satisfied, since there is just one line with two points on it. A3 is satisfied, again since there is just one line with two points on it. This is only one possible model; but there are many possible models that satisfy the axioms. A more complex one is shown in figure 4.4.

This could be implemented in a variety of ways (once we give an interpretation to the notions of "line" and "point" in terms of the implementation): on paper, on a transparency, using wood, nails, string, computer code, etc. So, we need a way of saying that "the nail" lines up with "point" (providing its meaning) and "the string" lines up with "line," in order to say that the structure (the system) is a model. A theory is then defined to be all of these possible models. In a nutshell, on the semantic view, a theory is a mathematical structure that models (describes, represents, etc.) the structure and behavior of a (target) system (the system the theory is supposed to capture). You can view a theory on this semantic conception as a family of abstract, idealized models of actual empirical systems. A correspondence needs to be established between the model and the empirical phenomenon, and this involves there being an "isomorphism" between the model and the phenomenon – an isomorphism just means that the structures are the same in some relevant sense, and so one can map one to the

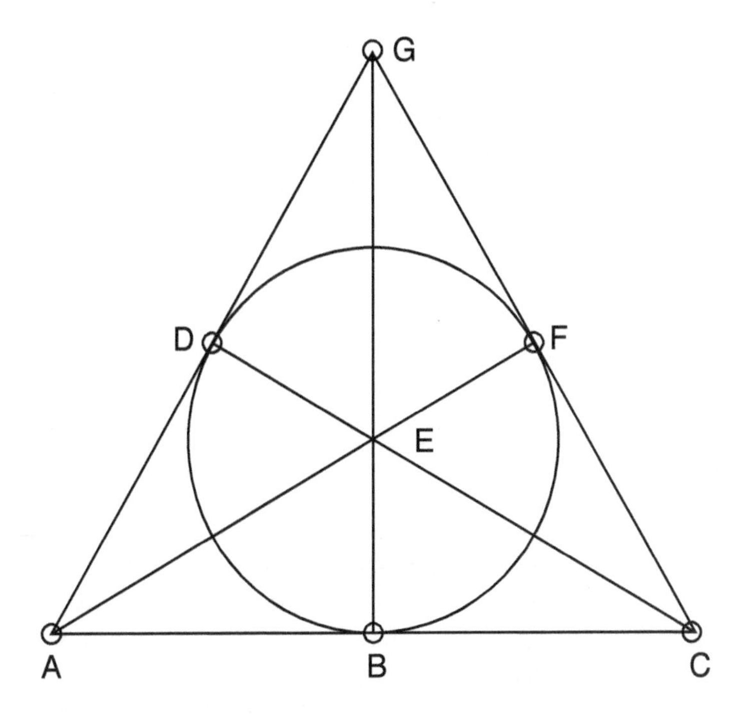

Figure 4.4 A model of axioms A1–A3: The system of points and lines makes the axioms true or, in more standard language, *satisfies* them. There are many other structures that could also satisfy the same axioms

Source: Bas van Fraassen, *Laws and Symmetry* (Oxford University Press, 1989), p. 219

other. Once we've established a correspondence between the model and some phenomenon, then explanation and prediction in the model constitutes explanation and prediction in the empirical world.

Let's put this slightly differently: we first must identify an "intended class" of systems that we are interested in, such as gases or liquids. We then present a formal structure (some mathematical entity like a geometry or vector space) and assert a mapping relation from this formal structure to the system of interest (e.g. from the geometry to the gas). This sets up a structural isomorphism (a correspondence) between the mathematical model and the thing being modeled. In this way, the semantic view is believed to provide a far better fit with the way scientists actually represent real-world systems. Let us now turn to the issue of representation of the world, and the question of whether our theories provide faithful representations of reality.

Representing and Realism

Much of our knowledge of climate change does not come from directly interacting with the world (with just looking at it), but from computer simulations. We trust these simulated worlds to tell us about the actual world. At the root of all of the vehicles of scientific knowledge is the notion of *representation*. While they might not look as obviously representational as the visual simulations in climate models, for example, theories too are often thought to represent systems in the world – of course, under the snazzy graphics of climate simulations, there lurk fairly abstract numbers and theory too that receive a graphical interpretation. The nice-looking climate simulations can be thought of as models in the above sense. Such models will use a whole bunch of theoretical notions in order to match observable phenomena, such as weather patterns (past and future). Does their success or failure mean that they are true of the world, so that these theoretical components have counterparts in the real world? This leads us into questions of reality.

An old debate concerns (metaphysical) "realism" *versus* "idealism": the world exists independently of human thought and perception *versus* the world is in some way dependent on the conscious activity of humans.

> The truth is "out there" *versus* the truth is "in here"!

Idealism can sound silly, but there are subtleties: it doesn't necessarily mean that there is no external world; it can mean that the external world is "conditioned" by our minds, as for example Kant believed, so that some features (colors or timbre, for example) are mind-dependent.

Our concern is with "scientific realism" *versus* "anti-realism" (or "instrumentalism"). Scientific realism is often presented as a view consisting of three key components:

- There is a world of objects, processes, and properties "out there," *independent* of us and our beliefs. Our statements about these things are true or false (or "approximately" so), and are *made so* by the things in the world (though we may never know if our statements are in fact true or false – still, there *is* a fact of the matter).
- The *aim* of science is to provide true descriptions of reality (though there can be other aims too: social advancement, for example).
- Science uses a variety of tools and methods (experiment, observation, statistics, etc.) for discovering the nature of reality. These methods are *fallible*, but they are the most reliable source of information we have.

Anti-realism is then presented as a view consisting of the following three contrary components:

- There is a world of *observable* objects, processes, and properties "out there," there *might* be a world of *unobservable* things too, but we can't know that, and it doesn't matter to science anyway. (According to some anti-realists, inasmuch as an "unobservable" part exists, it is the product of the human mind.)
- The *aim* of science is to provide true descriptions of a certain *part* of the world; namely, the "observable" part: science might well give a true account of the

unobservable part too, but that is of no consequence for the anti-realist.
- Science uses a variety of tools and methods (experiment, observation, statistics, etc.) for predicting and systematizing observable phenomena. These methods are *fallible*, but they are the most reliable source of information we have.

There is, then, no disagreement between realists and anti-realists about the observable part of reality: disagreement is posed at the level of the unobservable (or in logical empiricist terms, "theoretical") entities – anti-realists think that theoretical entities, such as "genes," "electrons," etc., are just convenient fictions; they help scientists make predictions of *observable phenomena* but should be given no ontological weight. The divergence concerns theories as providing true descriptions of reality *versus* theories as instruments for making predictions about observable phenomena. Obviously, realists are aware that theories are not always true, but they claim that, nonetheless, all theories are attempted descriptions of reality. Anti-realists disagree.

A common motivation for anti-realism is that scientific knowledge is limited by observation – fossils, birds, sugar crystals, and so on are all observable, and so absolutely unproblematic. Yet atoms, genes, quarks, and so on are not observable, which is a problem: how do we stretch our beliefs to them? This has some plausibility: at one stage scientists did doubt the existence of atoms, but no one (except a few heavily skeptical philosophers) has ever doubted the existence of birds. But *why* do scientists invoke such things as atoms and genes in the first place? The answer, according to anti-realists, is that they are *convenient*; they aid prediction. This leads to a crucial problem for anti-realists, as well as one of the primary motivations for realists. Theoretical entities, says the realist, might be convenient but this leaves open the issue of *how* they aid prediction if they do not exist! This problem forms the basis of the so-called "no miracles argument" for scientific realism:

> *"No Miracles Argument"*: The extraordinary predictive success of theories employing unobservable, theoretical entities (as a way of making predictions) warrants the belief in the existence of such entities: otherwise the (predictive) success of science would be a miracle!

So, the fact that our theories get things *right* (i.e. are "empirically successful," in that they make good predictions) coupled with the fact that they talk about unobservable entities (and these entities are *actively involved* in the making of such successful predictions), is *good evidence* for the hypothesis that those entities in fact exist – it is a form of "inference to the best explanation." Consider the laser, for example. This technology is based on "many-particle quantum mechanics." Basically, electrons in atoms go from higher to lower energy states and when they do, they emit photons (packets of light); but this emission triggers other electrons to emit photons, which trigger yet more emissions, resulting in a cascade. This leads to a very coherent light source. Clearly this description is highly theoretical: we are talking about electrons in atoms being stimulated to emit photons! None of this is directly observable. And yet lasers *work*: they let us get our shopping more quickly, correct faulty vision, play our CDs, and lots more. So the theory underlying the technology is empirically very successful and yet it involves extremely theoretical entities. Now, given this: wouldn't it be strange (a massive coincidence on a cosmic scale) if the theory of the laser made all these predictions while the entities involved in laser theory didn't exist? The coincidence would be staggeringly massively amplified by the fact that there are many theories that operate in a similar way – i.e.

laser theory isn't an isolated instance. How, if not by reifying the theoretical entities, are we to account for the close fit between theory and observation/experiment? Why, if there aren't such things as atoms, electrons, and photons, do lasers work?

The realist piggybacks on this successful deployment of theoretical entities using an inference to the best explanation format. That is, we are faced with a potentially puzzling fact: many theories that postulate theoretical (unobservable) entities are empirically very successful. If theories are true, and the entities in question really exist and behave according to theory's laws, then this success is far less of a puzzle. How could it be otherwise? If we reject this viewpoint, we are left with a serious puzzle that anti-realists cannot answer. This is the *positive* argument for realism: it is the only view that does not seem to make the success of science a miracle.

A *negative* argument for realism draws attention to the problems inherent in the "observable/unobservable" distinction: this really amounts to an attack on anti-realism, since any sensible anti-realist position will wish to say at least *some* things exist (tables, chairs, trees, etc.), and these will be observable things. If this distinction is central to anti-realist positions, and realism is the alternative to anti-realism, then an argument against this distinction is an argument *for* realism.

Why is this distinction a problem? Firstly, note that anti-realists take differing stances on scientific claims depending on whether they are about observable or unobservable things. They say: be, at best, agnostic about the latter but not the former. So, if this distinction between observable and unobservable things breaks down, then anti-realism is obviously in trouble. There are several problems with this distinction:

- *"Observation" and "Detection"*: is detection an instance of observation? For example, charged particles in "cloud chambers" leave tracks behind them as they interact with the chamber's contents (air and water vapor). These tracks are visible with the naked eye: does this mean electrons are observable? If yes, then the anti-realists are in trouble. Consider this though: jet aircraft can leave trails in the sky: does observing these trails count as observing the aircraft? Most people would say no: but surely *something* made those trails! The philosopher Ian Hacking argues that "if you can spray them, they exist!," meaning roughly that if you can actively manipulate something (if you can *create* tracks in a chamber) then the things making the tracks must exist (he calls this position "entity realism") – of course, this still does not mean we are observing them; only that we can *infer* their existence, which is more akin to detection than observation.

- *The "Observational Continuum"*: In his paper "The Ontological Status of Theoretical Entities," Grover Maxwell presents the following sequence of events, supposed to cause problems for the anti-realist:

  - Looking at something with the naked eye.
  - Looking at something through a window.
  - Looking at something through a pair of strong spectacles.
  - Looking at something through binoculars.
  - Looking at something through a low-powered microscope.
  - Looking at something through a high-powered microscope.
  - Continue in this same fashion ...

  Maxwell argued that these lie along a *smooth* continuum: so where is the cutoff between observable and unobservable? Does an astronomer merely *detect* the moon when looking through a telescope, much as physicists detect electrons using tracks? Just how sophisticated does scientific equipment have to be before it counts as detection rather than observation?

Maxwell argued that there is no principled way of answering such questions, and therefore anti-realism loses its classification of entities into observable ones and unobservable ones.

But the anti-realist Bas van Fraassen has a response: Maxwell has only shown that "observable" is a *vague* concept – it has borderline cases, where we aren't sure whether something is observable or unobservable, but there are also very clear-cut cases (the analogy is: "bald" is vague, but we can use the term perfectly well and say that a person with five hairs is bald whereas a person with a few million isn't). Hence, the distinction is vague but still perfectly usable: we can talk in a clear-cut fashion of chairs being observable and quarks being unobservable, despite the fact that cells, for example, sit somewhat uncomfortably between the two. However, unlike baldness, we seem to be constantly driving our observational equipment to greater resolutions, and coming up with new ways of observing what was once considered unobservable. It isn't clear whether the distinction is truly stable, or whether it is more of a hope that there are clear-cut cases that will stand the test of time.

- *The "Theory-Ladenness" of Observation*: We have met this problem already – Do a trained scientist and a lay person *see* the *same* thing when they look through a microscope at some cell? The trained scientist will surely be able to see certain areas as particular structures that perform some function: she will be able to see it *as* a cell. The lay person will no doubt see a blob! How far can we stretch this? How much of our observation is theory-laden? Even if a large part of it is, the anti-realist is in trouble for it shows that no clear-cut distinction can be made between the theoretical aspects of a theory, pertaining to electrons and genes and the like, and the observable aspects, such as observing a thermometer reading.

Problems for Realists

However, the anti-realists can fight back. They have two main weapons in their arsenal to try and destroy the realist position: the pessimistic meta-induction and the underdetermination arguments:

- *Pessimistic Meta-Induction (PMI)*: theories have been false many times in the past (despite enjoying empirical successes), and their theoretical ontologies (what they say exists and so what the realist is committed to) have been overturned. What makes realists so sure this won't happen again? Indeed, on the basis of experience, it is more likely than not to happen again.
- *Underdetermination of theory by data*: it seems like there can be "empirically equivalent" (i.e. matching in their observable predictions) theories with incompatible ontologies. We saw one example with Copernican and Ptolemaic theories that were based on the same data, but had radically different ontologies. Both can save the phenomena, but say very different things about the world. In this case, how is the no miracles argument supposed to get traction? Which ontology should we be led to?

PMI is a *historical* response: theories have been proven false and they will no doubt continue to be proven false in the future. Therefore, there is no foundation to realism: the ground will keep being swept from beneath believers in any theory's claims about the world beyond observations. Take as a common example the eighteenth-century "phlogiston theory of combustion" – when an object burns it releases a substance called *phlogiston*. This theory in its day was empirically successful: it fitted the data. A realist back then would have surely been committed to the existence of phlogiston. But Lavoisier showed it was false: burning occurs when things react with oxygen; all energy is conserved in the process. Realists have had the rug pulled from under them: the "stuff" they would have believed in was shown not to exist after all.

This argument is supposed to show that the no miracles argument is too quick: empirical success does not lead us automatically to the existence of theoretical entities after all. The anti-realists say we should therefore remain *agnostic* about the existence of the unobservable/theoretical content of our theories: it might be true, or it might not. Stick instead to what is stable over these changes in theory: the observable stuff.

Realists have responded to the argument in a variety of ways. They might say that empirical success of a theory is not evidence of the certain truth of what the theory says about the unobservable parts of reality; rather, it is evidence of the "approximate truth." This is a weaker notion and is supposed to be less vulnerable to PMI. The problem lies in making any kind of sense of "approximate truth": surely things are true or not? Another response is to say that "empirical success" should be defined not in terms of the fit between theory and known data, but in terms of the prediction of novel phenomena – phlogiston theory fitted the extant data but did not offer novel predictions. This response can perhaps serve to *reduce* the number of historical counterexamples, but it cannot eradicate them: the wave theory of light offers a good example that escapes both realist responses.

In 1690, Christiaan Huygens proposed a theory according to which light consisted of wave-like vibrations in an invisible, all-pervasive medium called *æther*. Newton advanced a rival theory according to which light consisted of particles. Wave theory was later accepted because of a remarkable (unexpected) prediction done on the basis of Augustin-Jean Fresnel's mathematical formulation of the theory. Poisson, critical of wave theory, deduced an observational consequence from the theory (i.e. a prediction) that he considered absurd ("a violation of common sense" no less): a bright spot should appear behind an opaque disc on which light is shone. But Dominique Arago verified the prediction! However, physics now tells us that there is no such thing as *æther*, and so the theory is incorrect despite its amazing prediction!

What is the moral of this tale? What this example shows is that a false theory can be extremely empirically successful, even to the point of making surprising predictions of novel phenomena. Moreover, how can a theory which talks about things that simply don't exist (namely *æther*) be even approximately true? Surely that would involve the things the theory talks about at least existing! Hence, this example kills even the modified versions of the no miracles argument. Anti-realists view this as showing that we cannot assume that empirical success implies even that modern scientific theories are *roughly* on the right track!

There is another realist modification that might allow the no miracles argument to be rescued: *structural realism*. According to this view, although our scientific theories often do indeed get the *nature* of things completely wrong (light being a wave rather than a particle, for example: though now we think of it as a particle *and* a wave!), they get the *structure* (or an aspect of the structure) of the world right. John Worrall, in his landmark paper "Structural Realism: The Best of Both Worlds?," argues that when theories are replaced, and old theoretical ontologies seemingly scrapped for new ones, there is nonetheless a continuity at the level of structure. In this way we both avoid the PMI (by finding something that is stable to be realist about) and respect the no miracles argument (by claiming that it is this stable structure that grounds the successes)!

What about the underdetermination argument? Both realists and anti-realists agree that theories are tested using observational consequences of theoretical entities. Of course, the anti-realists don't think of this as testing for the *existence* of those unobservable entities, but realists, as we have seen, do view this as support for their existence. So, all we have to go on is observable things: readouts on computer screens, dials, and so on. In a nutshell: observable data constitute the ultimate evidence for unobservable entities.

Take as an example the kinetic theory of gases (which we've mentioned several times now). This says that a sample of gas consists of molecules in motion. These molecules are unobservable: the realist will be committed to them, since they are part of the theoretical ontology of the theory; anti-realists will wish to remain agnostic about them. We clearly cannot test the theory by observing different gas samples, so we have to find some way of deriving observational consequences from the theory that can be tested directly. The theory predicts certain relationships between pressure and volume, so we can test these predictions, which can be tested directly: in a lab, isolate a gas sample, heat it up, and check for differences in the gas's volume using some apparatus. Now the problem is as follows: the observational data "underdetermine" the theories that scientists might produce. One and the same piece of observational data can be explained by multiple, theoretically incompatible, theories. In the case of gas theory: one possible explanation of the observational data is that gases consist of molecules in Newtonian motion (the kinetic theory of gases), but there may well be many other theories that can accommodate the same data: the gas might be conceived as a "fluid" obeying some appropriate hydrodynamic equation, for example, that is chosen to fit.

The problem here has a logical aspect that we have met several times before: a theory T might predict some observable effect O: $T \supset O$. However, just because we find O, this does not entail T: there might be many other theories T_n that imply *O* too. This is a little like the "curve-fitting" problem again: which itself was an instance of the problem of induction. So we have here a more sophisticated version of the problem of induction: the observational data do not entail the theory. So: how can the realist have any confidence in the truth of theories if there will always be competing theories that perform equally well as regards the observational data? Anti-realists say they cannot: score for them! This might be as simple as theories for the extinction of the dinosaurs: here there are several competing theories (meteorite strike, volcanoes, etc.) that all lead to the same outcome. Of course, in these cases there should be distinguishing traces from the theories that could be found in the historical record: craters formed at the appropriate time, for example. In this case, the realist would be within their rights to withhold belief.

However, the problem might be far worse: one might be able to construct "empirically equivalent" theories that will differ with respect to theoretical ontology, but that have *exactly* the same observable consequences for all possible observations. The anti-realist is fine here: one or the other theoretical ontology might be the genuine article, but we will never know. The realist, on the other hand, seemingly has to commit to one or the other: which one? The observational data cannot function as a guide here. Whatever theory is chosen, there will be an equally well-performing one with an incompatible ontology: agnosticism looks like the only option. You might say: "well, this all sounds terrible for the realist, but where are the examples?" It turns out there are many such examples, though coming from the bleeding edge of physics. The strangest example is probably the so-called AdS/CFT duality, in which a string theory with gravity on a five-dimensional space is equivalent in all observable/measurable respects to a quantum field theory without gravity on a four-dimensional space. We don't need to dwell on this. They are controversial, but they do show at least the possibility of genuine cases of underdetermination.

There are similar, more common cases in which we can give one and the same "theory" various distinct interpretations. You are probably aware of an example of this in the form of interpretations of quantum mechanics: many worlds versus consciousness-collapsing approaches, and many more. Einstein's theory of gravity (general relativity) provides another example: we can either view it in terms of a geometrical theory about the curvature of spacetime geometry, or we can view it as a theory of particles (gravitons) being exchanged on a totally flat spacetime. Again, there seems to be something transcending empirical success here, since they all share the same predictions. What is the realist to be realist about?

Again, however, there are realist responses:

- Often the realist will vehemently deny the genuineness of such empirically equivalent alternatives, claiming that they don't belong to mature or actual science. Underdetermination is a "philosopher's worry," they might say: the history of science does not show this competition between theories trying to account for the same data; it is hard enough finding *one* theory that does the job – this flies in the face of perfectly good examples, such as those just given.
- Even if they can be produced (and they really can!), they say that this does not mean the theories are *equally good*: one theory might be *simpler* than another, or be more intuitively plausible than another, or have fewer *types* of entity, or fewer independent assumptions, or might fit in with the rest of our scientific knowledge better. That is: there are non-empirical ways to break the apparent *impasse*. This is better, but the burden is then on the realist to justify whatever non-empirical principle is invoked to do the breaking.

- The structural realist approach can overlook the apparent differences between the underdetermined alternatives and point to their structural correspondence which they must share at some level in order to be empirically equivalent.

The anti-realist can deal with each of these: (1) such cases can be found, and if just one genuine case can be found then this is sufficient to pose a problem for the realist. (2) Why should theories that are "simpler," more parsimonious, more intuitive be seen as more likely to be true? True, empirically equivalent theories might be separable on these grounds, but these grounds are not reliable indicators of truth. Realism is about truth, and so these other factors must be shown to support this. (3) The structural realist needs to be careful to avoid the possibility that the only structural linkages that exist are at the level of empirical/observable structure, for then this becomes anti-realism in disguise (i.e. realist only at the level of observable stuff).

Another realist response turns the tables on the anti-realist so that the underdetermination argument is just as much of a problem for them: anti-realists believe in what can be observed. Note that this includes lots of things that haven't *actually* been observed: meteorites hitting the Earth and wiping out the dinosaurs for example. All the anti-realist has to go on are *traces* in what is observable now: we can look at various geological features – a big crater for example – and so on. But, the realist says, this puts the anti-realist in much the same position as realism with respect to the underdetermination argument: theories about unobserv*ed* objects are just as underdetermined as theories about unobserv*able* objects. This is borne out by the facts too: there are lots of competing theories about the extinction of the dinosaurs, involving meteors, volcanic eruptions that blocked out the Sun's radiation, and so on. In other words, consistent application of the underdetermination argument is just as destructive for the anti-realist for it implies that we can only have knowledge of things that have actually been observed (which is clearly very paltry and would rule out pretty much *all* of what we consider to be scientific knowledge).

The next step is to say that since science *does* give us reliable knowledge of the unobserved world, the underdetermination argument has to be wrong. Here we can see very clearly that we have another instance of the problem of induction: the inference from observed data to unobserved events and things or to unobservable events and things. The problem of induction remains of course, but we see that the underdetermination problem is not a *special* problem about unobservable entities; it poses as much of a problem for ordinary objects. Though we have come a long way since many of the topics covered in this small book, the problem of induction still lurks in the background of much current work, whether through the problems of observation, of underdetermination, or of evidence.

Summary of Key Points of Chapter 4

- The logical positivists were keen to nail down exactly what kind of thing a scientific theory was. Their answer was that it was a formal, logical structure of sorts, linked to the world by what they called "correspondence rules," associating theoretical entities (such as genes and atoms) with observable entities (such as marks on a computer screen). As with other aspects of the logical positivist position, this once popular view was heavily criticized and is now widely believed to be untenable for a variety of reasons, not least the difficulty of making sense of the division between theoretical and observable.
- An alternative response to the question of what a

scientific theory is is the "semantic view," which focuses not on the logical structure, but on the abstract entity represented by such structures: the *models*.

- The question of what a scientific theory is is related to the question of how it maps onto the world. The issue of whether our theories are then *true* or not and what this even means (aka "the scientific realism debate") then becomes central, with two broad classes emerging: realists who believe that the objects described by scientific theories really exist and those that do not (anti-realists), or do not believe that it matters (constructive empiricists).
- Much of the modern debate hinges on whether or not we take the success of science to be sufficient warrant for believing in the entities it postulates (the no miracles argument). The opposing arguments point to such things as the many revisions in science that appear to completely overturn belief in such entities (the pessimistic meta-induction argument) or the ability to construct equally successful theories that postulate different entities (the problem of underdetermination).

Further Readings

Books

- On the debate between realism and anti-realism, a particularly good overview, though a little dated, is Jack Smart's *Philosophy and Scientific Realism* (Routledge & Kegan Paul, 1963).
- A more recent, and sparklingly written, book on the same topic (staunchly defending realism) is James Robert Brown's *Smoke and Mirrors: How Science Reflects Reality* (Routledge, 1994).
- On the realism versus anti-realism debate, Bas van

Fraassen has two classic books: *Laws and Symmetry* (Clarendon, 1989) and *The Scientific Image* (Oxford University Press, 1980).
- Ian Hacking's entity realism, along with a rich discussion of many elements from this chapter, can be found in: *Representing and Intervening: Introductory Topics in the Philosophy of Natural Science* (Cambridge University Press, 2012).
- Stathis Psillos launches a defense of realism in: *Scientific Realism: How Science Tracks Truth* (Routledge, 1999).
- The best outline (including historical details) of the semantic view of theories is Frederick Suppe's: *The Semantic Conception of Theories and Scientific Realism* (University of Illinois Press, 1989).
- A book pushing the structural realist idea to its limit (so there are quite simply no objects!) is Steven French's *The Structure of the World: Metaphysics and Representation* (Oxford University Press, 2014).
- For a book on the extension of the realism debate into the "historical sciences" (such as paleobiology and geology, where the past rather than the tiny grounds the unobservable content), see Derek Turner's *Making Prehistory: Historical Science and the Scientific Realism Debate* (Cambridge University Press, 2014).

Articles

- Rudolf Carnap's own presentation of the syntactic view can be found in his 1956 paper "The Methodological Character of Theoretical Concepts," which is freely available to download from the University of Minnesota library: https://conservancy.umn.edu/handle/11299/184284.
- Hilary Putnam gives an exceptionally clear statement of the syntactic view, with a critique, in his paper

"What Theories are Not," in his collected papers, *Philosophical Papers, Volume 1: Mathematics, Matter and Method* (Cambridge University Press, 1975), pp. 215–27.
- Grover Maxwell's discussion of the "observation continuum" can be found in his "On the Ontological Status of Theoretical Entities," in H. Feigl and G. Maxwell (eds.), *Scientific Explanation, Space, and Time* (University of Minnesota Press, 1962), pp. 3–26.
- James Robert Brown provides a lucid treatment of the no miracles argument in "The Miracle of Science." *Philosophical Quarterly* (1982) **32**(128): 232–44.
- The classic statement of structural realism, as a way of navigating the pessimistic meta-induction and no miracles arguments, is John Worrall's "Structural Realism: The Best of Both Worlds?" *Dialectica* (1989) **43**(1–2): 99–124.
- André Kukla gives a nice discussion of the problems of underdetermination in "Does Every Theory Have Empirically Equivalent Rivals?" *Erkenntnis* (1996) **44**: 137–66.

Online Resources

As usual, there are numerous excellent entries from *The Stanford Encyclopedia of Philosophy* on topics relating to the present chapter:
- Holger Andreas's "Theoretical Terms in Science," *The Stanford Encyclopedia of Philosophy* (Fall 2017 Edition), E. N. Zalta (ed.): plato.stanford.edu/archives/fall2017/entries/theoretical-terms-science.
- Rasmus Winther's "The Structure of Scientific Theories," *The Stanford Encyclopedia of Philosophy* (Winter 2016 Edition), E. N. Zalta (ed.): plato.stanford.edu/archives/win2016/entries/structure-scientific-theories.
- Kyle Stanford's "Underdetermination of Scientific

Theory," *The Stanford Encyclopedia of Philosophy* (Winter 2017 Edition), E. N. Zalta (ed.): plato.stanford.edu/archives/win2017/entries/scientific-underdetermination.

- Anjan Chakravartty's "Scientific Realism," *The Stanford Encyclopedia of Philosophy* (Summer 2017 Edition), E. N. Zalta (ed.): plato.stanford.edu/archives/sum2017/entries/scientific-realism.
- Cosmologist Sean Carroll and philosophers of science Arthur Fine and Peter Dear debate the limits of science, including ideas relating to scientific realism in an episode of "Odyssey," on WBEZ Chicago Public Radio: youtube.com/watch?v=0qGcaJHF1Yg.
- The Rotman Institute of Philosophy has video of a wonderful talk by Bas van Fraassen, "The Semantic Approach to Science, After 50 Years": youtube.com/watch?v=6oM7-Wa_tAs. A Panel on Scientific Realism from a Rotman Institute Science and Reality Conference can be found at: youtube.com/watch?v=4gQ_7yIW2RQ.
- Michela Massimi has a nice selection of brief talks on scientific realism, the playlist for which is: youtube.com/watch?v=ywVtGOQxBW8&list=PLKuMaHOvHA4p0y6lBIGtOoke19DigM0gp.
- John Worrall has a nice statement on the problems of realism: youtube.com/watch?v=jm4H9nUsFpU.
- Paul Hoyningen-Huene has a good review on arguments against scientific realism here: youtube.com/watch?v=M6yVQETzecI.
- Finally, a very interesting set of talks from a lecture series (involving both scientists and philosophers of science) on Scientific Realism, organized by a group of students of the University of Vienna, can be found at the following playlist: youtube.com/playlist?list=PLtIs3eEC6pzL1v_haWfznvgiIqEK_dUo4.

Index